THE ECONOMICS AND ORGANIZATION OF BRAZILIAN AGRICULTURE

THE ECONOMICS AND ORGANIZATION OF BRAZILIAN AGRICULTURE

Recent Evolution and Productivity Gains

FABIO CHADDAD

AMSTERDAM • BOSTON • HEIDELBERG • LONDON
NEW YORK • OXFORD • PARIS • SAN DIEGO
SAN FRANCISCO • SINGAPORE • SYDNEY • TOKYO
Academic Press is an imprint of Elsevier

Academic Press is an imprint of Elsevier
125, London Wall, EC2Y 5AS.
525 B Street, Suite 1800, San Diego, CA 92101-4495, USA
225 Wyman Street, Waltham, MA 02451, USA
The Boulevard, Langford Lane, Kidlington, Oxford OX5 1GB, UK

Notices
Knowledge and best practice in this field are constantly changing. As new research and experience broaden
our understanding, changes in research methods, professional practices, or medical treatment may become
necessary.

Practitioners and researchers must always rely on their own experience and knowledge in evaluating and
using any information, methods, compounds, or experiments described herein. In using such information
or methods they should be mindful of their own safety and the safety of others, including parties for whom
they have a professional responsibility.

To the fullest extent of the law, neither the Publisher nor the authors, contributors, or editors, assume any
liability for any injury and/or damage to persons or property as a matter of products liability, negligence or
otherwise, or from any use or operation of any methods, products, instructions, or ideas contained in the
material herein.

British Library Cataloguing-in-Publication Data
A catalogue record for this book is available from the British Library

Library of Congress Cataloging-in-Publication Data
A catalog record for this book is available from the Library of Congress

ISBN: 978-0-12-801695-4

For Information on all Elsevier books,
visit our website at http://store.elsevier.com/

  **Working together
to grow libraries in
developing countries**

www.elsevier.com • www.bookaid.org

Publisher: Nikki Levy
Acquisition Editor: Nancy Maragioglio
Editorial Project Manager: Billie Jean Fernandez
Production Project Manager: Melissa Read
Designer: Maria Ines Cruz

Printed and bound in the United States of America

DEDICATION

To Maria and Rodrigo.

CONTENTS

PREFACE

It was February 2013, in the middle of the summer season in Brazil, when most Brazilians take some time off to go to the beach. In particular, the *paulistanos* — the residents of the largest city in the country, São Paulo — love to escape the city in the weekends and drive about 100 km down Serra do Mar to the coast. But the roads were more clogged than usual. A strike in the Port of Santos resulted in a long line of trucks blocking access to the beaches. Most of these trucks were carrying agricultural commodities and food products to be exported to world markets. The *paulistanos* were furious they could not get to the beach. At the same time, I was working with producer groups in Mato Grosso, about 2,000 km to the northwest of the port. They were concerned about the delays the strike was causing. The Chinese — the major importers of soybeans from Brazil — were cancelling orders as a result of the delays and were sourcing product from competitors. One of the farmers complained to me that they had been calling the attention of the Brazilian government to the logistical nightmare of moving farm commodities from the cerrado region to the ports in the southeast, but nothing had been done about it. "At least now they will pay attention to our problem and invest in transportation infrastructure to move grains from Mato Grosso to the ports in the northern and northeastern region and avoid clogging the already congested infrastructure in the south. This will not happen because it makes economic sense; it is just to make sure the *paulistanos* can get to the beach." This incident highlights the big disconnect that exists between politicians, urbanites, and farmers in Brazil.

In April 2014, the *Economist* published an article titled "The 50-year snooze: Brazilian workers are gloriously unproductive. For the economy to grow, they must snap out of the stupor." The data about productivity in Brazil are appallingly bad. The article goes on to say that "apart from a brief spurt in the 1960s and 1970s, output per worker has either slipped or stagnated over the past half century, in contrast to most other big emerging economies. Total factor productivity, which gauges the efficiency with which both capital and labor are used, is lower now than it was in 1960. If the economy is to grow any faster than its current pace of 2% or so a year, Brazilians will need to become more productive." In other words, the low productivity of the Brazilian economy is *the* major constraint for the country's development.

But there is one sector of the Brazilian economy that is a world leader in productivity gains in the last 40 years — agriculture. As a result of total factor productivity gains of 3.0% per year since the 1970s, Brazil has become the largest net food exporter in the world. But the majority of Brazilians — including many well-known economists — are not aware of this success story. In fact, many Brazilians still believe that agriculture is a backward and inefficient sector of the economy with large, unproductive farms (*latifúndios*) coexisting with millions of poor smallholders. Interestingly enough, many outside Brazil are aware of the agricultural transformation that the country has gone through in the last four decades and believe that Brazilian agriculture offers a solution to food security and rural poverty issues in Africa. Natural conditions — soil and climate, in particular — are similar and thus the tropical agricultural technologies developed by Brazilian researchers can help increase farm productivity in the African savannahs. But agricultural development is a complex process, which requires entrepreneurship and organization to come to fruition.

This book analyzes how Brazilian agriculture has changed since the 1970s and how the country became a top-five producer of 36 agricultural commodities globally. Chapter 1 describes the evolution of Brazilian agriculture since 1970, focusing on production growth, the increased use of modern farm inputs, productivity gains, and the economic effects of what is largely a success story. Chapter 2 discusses the *enabling conditions* of such production and productivity gains, including natural resource availability, the development of agricultural technologies adapted to tropical conditions, and changes in agricultural policy. These enabling conditions are necessary but not sufficient for productivity gains to occur, which begs the question of *how* Brazilian farmers overcame many challenges to adopt technology, increase production, and gain international competitiveness.

The aim of this book is to call attention to two oft-neglected factors associated with agricultural development — entrepreneurship and value chain organization. Starting in Chapter 3, we analyze how different value chain configurations provided technology and credit to Brazilian farmers and linked them to domestic and global markets. Chapter 3 focuses on the organization of agricultural value chains in the southern region where cooperatives and contract farming arrangements play a prominent role in linking farmers to markets. Chapter 4 analyzes productivity gains and increased competitiveness of the sugarcane and orange juice sectors in the southeastern region and highlights the role of vertically integrated agribusiness and producer organizations in value chain organization.

Chapter 5 describes how the cerrado has been developed since the 1970s and has become a breadbasket. The chapter explains the role of public— private partnerships in developing technologies adapted to the cerrado conditions and how private colonization firms and agricultural cooperatives played a crucial role in settling pioneers and provided them with the technologies and services necessary to farm in the agricultural frontier. The chapter provides examples of how pioneers were able to survive several economic crises and become commercial producers in the frontier, with a focus on entrepreneurship, economies of scale, and the emergence of new-generation cooperatives. Chapter 6 summarizes the main findings of the book and emphasizes the roles of entrepreneurship and value chain organization in agricultural development. It also provides some perspectives on progress being made to ameliorate environmental and social concerns associated with agricultural development in Brazil.

This book is based on many case studies, vignettes, and stories to show with micro-analytic detail how farm and agribusiness entrepreneurs overcame many challenges to build world class organizations and effective supply chains that provided the basis for productivity gains and increased international competitiveness. These case studies, vignettes, and stories were developed from personal interviews conducted in 2014 with farmers, agribusiness managers, industry leaders, policymakers, and experts.

I would like to thank the following individuals who agreed to share their knowledge and wisdom with me:

- José Garcia Gasques (MAPA)
- Maurício Lopes, Elísio Contini, and Geraldo Martha (EMPRAPA)
- João Veloso Silva, Lineu Domit, Austerclínio Farias Neto, Marcelo Carauta, and Júlio Reis (EMBRAPA Agrossilvipastoril)
- José Américo Rodrigues (ABRASEM)
- Rodrigo Santos and Geraldo Berger (Monsanto do Brasil)
- Ivo Carraro (Coodetec)
- Jorge Karl, Manfred Majowski, Norbert Geier, Adam Stemmer, and Arnaldo Stock (Agrária)
- Frans Borg (Castrolanda)
- Daniel Dias (Coonagro)
- Marcos Jank (BRF)
- Eduardo Leão de Sousa, Adhemar Altieri, Geraldine Kutas, and Maria Luiza Barbosa (UNICA)
- Luís Pogetti and Soren Jensen (Copersucar)
- Carlos Dinucci (Usina São Manuel)

- Luiz Antonio Dias Paes (CTC)
- Murilo Parada (Louis Dreyfus Commodities)
- Décio Tocantins (AMPA)
- Luciane Copetti (Prefeitura de Lucas do Rio Verde, MT)
- Ricardo Tomszyk, Marcelo Duarte, Nery Ribas, Cid Sanches, and Susiane Azevedo (Aprosoja)
- Silvésio de Oliveira (farmer in Tapurah, MT)
- Nelson Piccoli (farmer in Sorriso, MT)
- Francisco Soares Neto (TMG and FMT)
- Rodrigo Rodrigues and Fabiano Costa (Agrifirma)
- Júlio Piza (BrazilAgro)
- Gilson Pinesso and Ademir Pinesso (Produzir S.A.)
- Eraí Maggi (Grupo Bom Futuro)
- Adair Mazzotti (OCB-MT)
- José Carlos Dolphini, Alexandre Bottan, and Carlos Menegati (Cooperfibra)
- Gilberto Peruzzi and Evandro Lermen (Coacen)
- Otávio Palmeira, João Luiz Ribas Pessa, and Adelar Dahmer (Unicotton)
- Helvio Fiedler (Coabra)
- João Vianna (IGEAGRO)
- Luiz Fernando do Amaral (Rabobank do Brasil)
- Marcelo Pereira de Carvalho (Agripoint)
- Christiano Nascif (Labor Rural)

I would also like to thank Jill Findeis (Division of Applied Social Sciences, University of Missouri) and Marcos Lisboa (INSPER) for encouraging me to write this book and providing the necessary time and resources to conduct the field research in 2014. Special gratitude is due to my dear friends Sérgio Lazzarini (INSPER) and Henrique Americano de Freitas (Minerva Foods), who read, critiqued, and provided helpful comments in earlier versions of the chapters. I could not have finished this book without your intellectual input and constant encouragement. I share the qualities of this book with these people but I retain full responsibility for its shortcomings and errors.

Fabio Chaddad
Associate Professor,
University of Missouri and INSPER,
Columbia, MO

CHAPTER 1

Introduction

Contents

In 1923, Ferruccio Pinesso arrived in the port of Santos, Brazil, accompanied by his wife, Anna, his mother, and his five brothers and sisters. The Pinesso family decided to leave poverty and famine in Italy to try their luck in the New World. At that time the Brazilian economy depended heavily on coffee exports and most immigrants found work in coffee plantations in two states — São Paulo (SP) and Paraná (PR). They lived on the farm, took care of the coffee trees, and were paid a share of the crop — typically 30—40% — as remuneration for their work. Depending on the arrangement with the landlord, the family could use the land around the household to produce food for its own subsistence.

And so did the Pinesso family. The family first established itself in Marcondésia, SP, located 420 km northwest of the state capital and then moved to northern Paraná. When Ferruccio died in 1951, he left his wife and seven children a small plot of land that he had acquired in 1947. As was common among Italian immigrants, the oldest son would be responsible for taking care of the family. And so was the challenge of Eugênio Pinesso, who was 22 years old when his father passed away. He sold the small plot of land that his father had bought and with savings from a successful coffee crop and a loan from his brother-in-law was able to acquire a small farm in Peabiru, PR. Even though he had not been able to go to school as a child, Eugênio was a hard-working, entrepreneurial, and shrewd businessman. He was an early adopter of new agricultural practices and technologies, such as using coffee straw as a source of natural

F. Chaddad: The Economics and Organization of Brazilian Agriculture.
DOI: http://dx.doi.org/10.1016/B978-0-12-801695-4.00001-X

1

fertilization, coffee shading, liming to improve soil fertility, early planting of beans (to sell in the off-season), and new farm machinery.

He was also one of the first farmers to plant soybeans as a summer crop in Paraná in the late 1950s, when it was a specialty crop known as "Japanese bean." First, Eugênio planted soybeans between coffee tree rows to fixate nitrogen in the soil and increase coffee yields. With the discovery of soil liming and the use of chemical fertilizers, Eugênio started to grow soybeans as the main crop in regions of the state with low natural soil fertility, where coffee plantations were not viable. By 1972, Eugênio was a well-established farmer with six farms in Paraná, totaling 1,500 hectares (ha), and a retail business.

Following the 1975 frost that decimated millions of coffee trees in Paraná, Eugênio decided to acquire his first farm in the cerrado — 2,000 ha in the state of Mato Grosso do Sul (MS) in 1976. He experimented with several crops and was convinced that the cerrado soils could be as productive as in Paraná. In 1983 Eugênio took the bold step of selling all his land in Paraná to acquire more land in Mato Grosso, and the family moved to Campo Grande, MS. "Farmers received incentives from the government to develop the land and credit to buy machinery, fertilizers and lime. Dirt roads received asphalt to allow the crops to be transported to markets. I told my friends in Paraná that if they wanted to be real farmers — rather than peasants or hobby farmers — they needed to move to Mato Grosso."[1] In 2014, the Pinesso Group planted 117,000 ha in the Brazilian cerrado and had more than 1,000 employees. We will see later in this book how the Pinesso Group has expanded since the 1980s to become one of the major farming entities in the Brazilian cerrado.

As a result of the entrepreneurship of farmers like Mr. Pinesso, Brazilian agriculture has experienced significant growth in the last four decades. Between 1975 and 2010, total agricultural production in Brazil grew fourfold, with an annual average growth rate of 3.7%, making it a top-five producer of 36 commodities globally by 2008 (for details see Rada and Buccola (2012)). During the same period, total factor productivity (TFP) growth averaged between 3.0 and 3.4% per year. The substantial agricultural production growth is, therefore, mainly attributed to increased productivity of Brazilian farmers. As a result of productivity gains, Brazil was able to achieve food security, real food prices decreased, households spent a decreasing share of their income on food, and Brazil became one of the main agricultural producers and exporters globally.

How did Brazil become an agricultural powerhouse? Most economists who have studied Brazilian agriculture would agree that this success story is largely due to natural resource availability — in particular, land, water, and favorable tropical climate — public and private investments in agricultural technologies adapted to the tropics, and changes in agricultural policy since the 1970s. Because of these enabling conditions, Brazilian farmers were able to take advantage of the 2000s commodity boom and prosper. This book discusses these factors, but adds another dimension to the analysis — that is, a micro-level look at the organization, governance, and strategic changes adopted by farm and agribusiness entrepreneurs that are the ultimate cause of productivity gains. As the Pinesso story shows, it is the rural entrepreneur in the farm, cooperative, or agribusiness enterprise who bears the risks of Mother Nature and commodity markets, who adopts technology and who makes strategic decisions about what, where, and how to produce. He or she makes productivity gains a reality. But before looking at these micro-level changes, let's first take a closer look at the numbers.

1.1 PRODUCTION

Simply stated, TFP is the ratio of an aggregate index of output to an aggregate index of inputs used in production. To be able to measure productivity and estimate productivity growth over time, the first necessary step is to literally add "apples to oranges" to get an aggregate measure of production. Obviously, there are serious methodological issues in aggregating apples, oranges, and all other farm products in a single, aggregate measure. I will not explain these technical challenges here but refer the interested reader to the seminal work of agricultural economist Bruce Gardner (2002). Rather, I will rely on data from two recent attempts to estimate agricultural productivity — one in Brazil (Gasques et al., 2014) and the other internationally (Fuglie et al., 2012). Gasques et al. (2014) use municipal agricultural production (PAM) and municipal livestock production (PPM) data collected by the Brazilian Institute of Geography and Statistics (IBGE) and agricultural price data from Fundação Getúlio Vargas (FGV) to construct their aggregate measure of farm output, which includes 66 temporary and permanent crops and 11 livestock products. The farm output index of Fuglie et al. (2012) is the sum of the production value of 189 crop and livestock commodities, valued at constant, global-average prices from 2004–2006. Their data are from the Food and Agriculture Organization of the United Nations (FAO).

Although these sources use different production and price data to estimate an index of farm output, their results are strikingly similar. Both estimates indicate a fourfold increase in agricultural production value between 1975 and 2010, which is equivalent to a growth rate of 3.7% per year (Figure 1.1). In constant 2005-dollar terms, the total value of

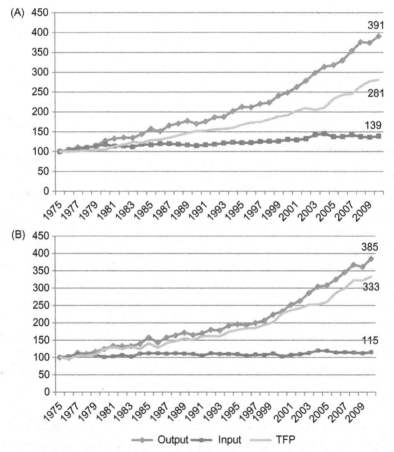

Figure 1.1 Production growth and productivity gains in Brazilian agriculture (1975–2010). *Sources: (A) Fuglie et al. (2012) and (B) Gasques et al. (2014).*

Brazilian farm output increased from \$35.8 billion in 1975 to \$140.1 billion in 2010. During this period, grain production increased from 39.4 to 149.2 million tons and meat production increased from 3.4 to 24.6 million tons. Substantial production growth was also observed in sugarcane, coffee, milk, fruits, and vegetables. Table 1.1 shows the evolution of Brazilian agriculture in numbers since 1975.

Table 1.1 Evolution of Brazilian agriculture in numbers (1975–2010)

	1975	1980	1990	2000	2010
Total value of production (US\$ million)[a]	35,883	45,839	61,052	89,187	140,127
Grain production (million metric tons)[b]	39,400	50,871	58,280	83,030	149,255
Meat production (million metric tons)[c]	3,394	5,385	8,414	14,510	24,622
Agricultural land (1,000 ha)[d]	43,387	52,998	56,230	66,362	91,426
Grain harvested area (1,000 ha)[e]	32,900	40,158	38,945	37,824	47,416
Agricultural labor (1000 persons)[f]	15,760	16,342	14,062	13,325	11,049
Animal capital stock (1,000 head)[g]	122,199	150,459	184,002	207,220	254,819
Farm machinery capital stock (number of tractors)[h]	338,613	563,205	751,779	824,466	815,053
Fertilizer consumption (1,000 metric tons)[i]	2,190	4,635	3,460	7,807	10,772

[a]Sum of the value of production of 189 crop and livestock commodities, valued at constant, global-average prices from 2004–2006 and measured in constant 2005 US dollars. Source: Estimated by Fuglie et al. (2012) with FAO data.
[b]Source: CONAB (2014).
[c]Includes beef, veal, poultry, and pork in million metric tons of carcass-weight equivalents. Source: USDA, PSD online database (2014).
[d]Rainfed cropland equivalents (quality-adjusted irrigated, rainfed cropland, and permanent pasture). Source: Estimated by Fuglie et al. (2012) with FAO data.
[e]Source: CONAB (2014).
[f]Economically active persons employed in agriculture (male and female). Source: Estimated by Fuglie et al. (2012) with FAO data.
[g]Number of cattle-equivalent head of livestock on farms. Source: Estimated by Fuglie et al. (2012) with FAO data.
[h]Number of 40-CV tractor-equivalent machinery units in use. Source: Estimated by Fuglie et al. (2012) with FAO data.
[i]Sum of N, P_2O_5, and K_2O fertilizers in tons of N-fertilizer equivalents. Source: Estimated by Fuglie et al. (2012) with FAO data.

1.2 FARM INPUTS

A second step in measuring agricultural productivity is to estimate an aggregate of inputs used in production, including land, labor, capital, and other farm inputs such as seeds, fertilizers, crop protectants, and animal feed. Gasques et al. (2014) use land data collected by IBGE (Agricultural Census and PAM), which include crop-harvested area and pastures, and land rental price data from FGV. Labor is estimated from the number of workers employed in agriculture and salaries obtained from the National Household Survey Research (PNAD). The number of farm machines used in agriculture is obtained from the National Association of Automobile Manufacturers (ANFAVEA) and prices are estimated from manufacturers' financial statements. Annual machinery use is calculated with the assumption that farm machinery depreciates in 16 years. Fertilizer and crop protectant use is obtained from industry sources such as ANDA, Potafos, and SINDIVEG. The farm input data are aggregated without any quality adjustments. Fuglie et al. (2012) use FAO data to estimate the total input index, where land, capital, and NPK fertilizers are quality-adjusted. It also includes the number of economically active adults in agriculture (labor), total livestock capital on farms in "cattle equivalents", and total animal feed from crops and crop-processing residues.

Not surprisingly, the estimated input indexes in Gasques et al. (2014) and Fuglie et al. (2012) are markedly different. The Gasques et al. (2014) data suggest farm input use increased 15% or 0.26% annually between 1975 and 2010, whereas the Fuglie et al. (2012) data indicate a higher input growth rate of 39%, equivalent to 0.82% per year, in the same period. The major discrepancy between the two studies is the estimated land use in agriculture. Gasques et al. (2014) estimate a meager 2% land use increase during the period. The quality-adjusted land use index estimate in Fuglie et al. (2012) more than doubled from 43.3 million hectares in 1975 to 91.4 in 2010 (Table 1.1). Both studies, however, report a significant reduction in the number of workers employed in agriculture − between 18% (Gasques et al., 2014) and 30% (Fuglie et al., 2012). According to the FAO data, the number of economically active workers employed in Brazilian agriculture decreased from 15.7 million in 1975 to 11 million in 2010. Both studies show increases in farm machinery capital stock and consumption of raw materials such as fertilizers and crop protectants. The FAO data show that farm machinery capital stock increased from 338,000 40-cv tractor-equivalents in 1975 to 815,000 in 2010. During the same time, the use of NPK fertilizers increased from 2.2 to 10.8 million tons of N-fertilizer equivalents.

1.3 PRODUCTIVITY GAINS

As a result of different estimates for input use in agriculture, the resulting productivity indexes and growth estimates are not the same (Figure 1.1). TFP growth in Brazilian agriculture between 1975 and 2010 was 3.41% per year in Gasques et al. (2014) and 2.96% in Fuglie et al. (2012). This is not surprising given that both studies use different data sources and methodologies to aggregate the data. But the common conclusion is that the growth of Brazilian agricultural production in the last four decades can be primarily explained by increased productivity. The results from both studies also suggest that productivity growth has accelerated to 4% annually in the last decade (Table 1.2).

How does the performance of Brazilian agriculture compare with other countries? Table 1.3 compares agricultural production growth and productivity gains among the top-20 agricultural producers. Between 1971 and 2010, annual agricultural production growth in Brazil averaged 3.78%, in line with most developing countries in the table and significantly higher than developed countries. Brazilian agricultural production growth accelerated in the 2000s to 4.36% per year, second only to Indonesia (4.51%). The relative performance of Brazilian agriculture was even better considering TFP growth — 2.79% per year in the 1971—2010 period (second only to Spain with 2.84%) and accelerating to 4.02% annually between 2001 and 2010 (second only to Italy with 4.41%).

Table 1.4 shows data that allow for a broader perspective on agricultural productivity gains across the world in the 2000s. Total global

Table 1.2 Productivity growth in Brazilian agriculture (1975—2010)

	Annual growth rates (%) Fuglie et al. (2012)			Annual growth rates (%) Gasques et al. (2014)		
	Output	Input	TFP[a]	Output	Input	TFP[a]
1975—1980	4.32	3.22	1.04	4.54	0.42	4.16
1981—1990	3.43	0.37	3.07	3.02	0.80	2.22
1991—2000	3.66	1.03	2.61	3.17	−0.28	3.47
2001—2010	4.36	0.35	4.02	4.69	0.43	4.21
1975—2010	3.78	0.82	2.96	3.68	0.26	3.41

[a]Total factor productivity.

Table 1.3 Productivity gains among largest agricultural producers (1971–2010)

	Average annual production level (US$ million)[a]		Average annual production growth (%)		Average annual total factor productivity growth (%)	
	1971–1980	2001–2010	1971–2010	2001–2010	1971–2010	2001–2010
China	120,453	465,063	4.54	3.46	2.58	2.93
United States	147,448	224,875	1.44	1.39	1.78	2.05
India	81,552	194,409	2.92	3.70	1.27	2.02
Brazil	37,241	118,234	3.78	4.36	2.79	4.02
Indonesia	15,837	49,737	3.77	4.51	1.50	3.12
Russian Federation	60,831	48,532	−0.91	0.94	1.08	3.13
France	36,948	42,339	0.46	−0.62	2.38	2.40
Argentina	19,860	37,499	2.13	3.09	0.77	1.45
Germany	34,334	35,384	0.01	0.72	2.75	3.02
Pakistan	11,595	33,810	3.62	3.24	1.47	0.25
Mexico	15,717	33,702	2.50	1.72	1.13	2.02
Spain	19,742	33,166	1.72	−0.02	2.84	2.99
Turkey	18,254	33,116	2.01	1.82	1.43	1.86
Italy	29,355	32,571	0.35	0.08	2.41	4.41
Nigeria	9,789	32,260	4.14	1.31	2.01	−0.15
Thailand	11,683	28,406	2.97	1.89	1.86	1.70
Canada	15,200	26,730	1.90	2.24	2.12	1.91
Australia	15,307	24,560	1.68	−0.64	1.47	0.53
Vietnam	5,749	24,177	4.72	3.91	2.22	2.20
Iran	6,535	24,155	4.37	2.74	2.43	1.31

[a]Gross agricultural production value in constant 2005 US dollars.
Source: Estimated by Fuglie et al. (2012) with FAO data.

Table 1.4 Agricultural productivity gains across the world (2001–2010)

	Output (%)[a]	Input (%)[b]	TFP (%)[c]
World	2.50	0.70	1.81
Developed countries	0.59	−1.73	2.32
North America	1.33	−0.77	2.10
Europe	−0.13	−2.20	2.07
Transition economies	1.49	−0.41	1.90
Developing countries	3.39	1.20	2.20
East and South Asia	3.40	0.71	2.69
Sub-Saharan Africa	3.26	2.28	0.99
West Asia and North Africa	2.42	0.39	2.04
Latin America	3.37	0.70	2.67
Brazil	4.36	0.35	4.02

[a]Average annual growth in gross agricultural production value.
[b]Average annual growth in farm inputs used in production.
[c]Average annual growth in total factor productivity.
Source: Estimated by Fuglie et al. (2012) with FAO data.

agricultural output grew 2.5% on average per year between 2001 and 2010, with 3.39% in developing countries, 1.49% in transition economies, and 0.59% in developed countries. At an average annual rate of 4.36%, Brazil achieved significantly faster production growth than its developing country peers. In the same decade, TFP growth averaged 1.81% annually across the globe, with 2.32% in developed countries, 2.2% in developing countries, and 1.9% in transition economies. With 4.02% annual TFP growth, productivity gains in Brazilian agriculture were significantly higher than these global and regional averages.

The outstanding relative performance of Brazilian agriculture is noteworthy because the country's economy as a whole has not performed well in the last four decades. Productivity in Brazil is low − even compared to other developing countries − and has stagnated since the 1980s. Apart from the Brazilian "economic miracle" of the 1970s and a brief spurt in the mid- to late-2000s, TFP in the Brazilian economy has not grown much (Table 1.5). The low and relatively stagnant productivity in the Brazilian economy is attributed to many factors, including high costs of doing business, high tax rates and import tariffs, inadequate infrastructure, low public and private investments in R&D, and low levels of education. The significant growth and productivity gains in Brazilian agriculture described above occurred despite all these challenges that affected all sectors of the economy.

Table 1.5 Total factor productivity (TFP) growth in the Brazilian economy (1970–2010)

	TFP growth (%) Brazil[a]	TFP growth (%) agriculture[b]
1971–1980	2.0	N/A
1981–1990	−1.2	2.22
1991–2000	0.3	3.47
2001–2012	0.8	4.06

[a]Source: Bonelli (2013).
[b]Source: Gasques et al. (2014).

1.4 ECONOMIC EFFECTS

As it was spurred by productivity gains, the recent growth of Brazilian agriculture has generated many positive effects. It means that more output — food — is produced with fewer resources. Of course, agricultural development has not occurred without environmental and social impacts, but the broad picture paints what is largely a success story. First and foremost, Brazil was able to achieve food security. Until the late 1970s, Brazil was a net importer of many food products and received food aid from abroad. Since then, food availability has increased significantly. Grain production increased from 342 kg per capita in 1975 to 761 kg in 2010, while meat production increased almost fourfold from 28 to 105 kg per capita and milk production per capita more than doubled from 74 to 161 liters. As a result of increased food availability, the domestic price of food decreased at an annual average rate of 2.2% in real terms between 1976 and 2012. The price of the basic food basket in 2012 was 79% lower in real terms than in 1976 (for details refer to Alves et al. (2013)). According to the Family Budget Survey (POF) data collected by IBGE, a typical Brazilian family spent 34% of the household budget on food in 1975, which decreased to 16% in 2009.

The second major positive effect was that Brazil became a leading agricultural exporter. According to World Trade Organization (WTO) data, Brazilian agricultural exports grew 6.9% annually in real terms since 1980, from US$10.1 billion to US$86.4 billion in 2012. Brazil is now the third largest agricultural exporter in the world behind the European Union and the United States with a 5.2% share of world agricultural exports (Table 1.6). Given that Brazil imported US$13.1 billion in agricultural products in 2012, it is the largest *net* exporter with a surplus of US$73.3 billion. Net agricultural exports generated the accumulated sum of US$664 billion to the

Table 1.6 Leading exporters and importers of agricultural products[a] (2012)

	Value (US$ billion) 2012	Share of world exports/ imports (%)				Annual growth (%) 2005–2012
		1980	1990	2000	2012	
Exporters						
European Union (27)	613	–	–	41.8	37.0	7
Extra-EU exports	163	–	–	10.2	9.8	10
United States	172	17.0	14.3	13.0	10.4	11
Brazil	86	3.4	2.4	2.8	5.2	14
China	66	1.5	2.4	3.0	4.0	13
Canada	63	5.0	5.4	6.3	3.8	6
Indonesia	45	1.6	1.0	1.4	2.7	18
Argentina	43	1.9	1.8	2.2	2.6	12
India	42	1.0	0.8	1.1	2.6	22
Thailand	42	1.2	1.9	2.2	2.5	13
Australia	38	3.3	2.9	3.0	2.3	9
Malaysia	34	2.0	1.8	1.5	2.0	14
Russian Federation	32	–	–	1.4	1.9	12
Vietnam	25	–	–	0.7	1.5	19
New Zealand	24	1.3	1.4	1.4	1.4	9
Mexico	23	0.8	0.8	1.7	1.4	9
Above 15	**1,349**	–	–	**83.3**	**81.4**	–
Importers						
European Union (27)	623	–	–	42.6	35.7	6
Extra-EU imports	173	–	–	13.3	9.9	6
China	157	2.1	1.8	3.3	9.0	19
United States	142	8.7	9.0	11.6	8.1	6
Japan	94	9.6	11.5	10.4	5.4	5
Russian Federation	42	–	–	1.3	2.4	14
Canada	38	1.8	2.0	2.6	2.2	8
Republic of Korea	33	1.5	2.2	2.2	1.9	10
Saudi Arabia	29	1.5	0.8	0.9	1.7	18
Mexico	27	1.2	1.2	1.8	1.6	7
India	26	0.5	0.4	0.7	1.5	19
Hong Kong, China	25	–	–	–	–	12
Retained imports	17	1.0	1.0	1.1	1.0	13
Malaysia	21	0.5	0.5	0.8	1.2	17
Indonesia	21	0.6	0.5	1.0	1.2	16
Egypt	18	0.6	1.1	0.7	1.0	20
Thailand	17	0.3	0.7	0.8	1.0	13
Above 15	**1,304**	–	–	**81.6**	**74.8**	–
Brazil	13	0.9	0.6	0.8	0.7	16

[a]Agricultural products include SITC sections 0, 1, 2, 4 minus 27 and 28.
Source: World Trade Organization (2014).

Brazilian merchandise trade balance between 1980 and 2012. This surplus in agricultural trade was more than enough to offset an accumulated merchandise trade deficit of US$313 billion since 1980, which includes manufactures, fuels, and mining products (Figure 1.2).

In addition, the fact that Brazil was able to generate increasing food surpluses helped to alleviate food security issues worldwide. WTO data show that world food imports increased 5.7% annually from US$233 to US$1,468 billion between 1980 and 2012 as a result of population and income growth and urbanization, primarily in developing countries. As this trend continues in the next decade, substantial production expansion will be necessary to meet the needs of a growing world population. The expected additional production includes 17% more coarse grains, 26% more oilseeds, 27% more poultry, and 20% more sugar by 2023 relative to the base period of 2011−2013. This production growth will come primarily from developing countries in Asia and Latin America, with Brazil expected to strengthen its position as a major net exporter (OECD-FAO, 2014).

1.5 SKETCH OF THE BOOK

The objective of this book is to describe how, since the 1970s, Brazil has become one of the largest producers and exporters of agricultural products. As explained above, the major reason behind this phenomenon was productivity growth of 3.0−3.4% per year achieved by Brazilian farmers between 1975 and 2010. Productivity growth, in turn, can be largely explained by a set of enabling conditions, which include natural resource availability and agricultural policies that evolved since the 1960s to provide the needed resources for farmers to transform land, water, and solar energy into agricultural and food products. These resources include agricultural technology adapted to the tropics and rural credit. Macroeconomic stability, increasing openness to trade and deregulation in the 1990s exposed Brazilian farmers to international markets, thereby providing strong economic incentives for productivity gains. The 2000s commodity boom further spurred the entrepreneurial spirits of farmers to expand production and use modern technologies. Chapter 2 analyzes these enabling conditions and the policy environment that fostered productivity growth in Brazilian agriculture.

There is nothing novel about these explanations and most economists appear to be content with them. These enabling conditions are necessary

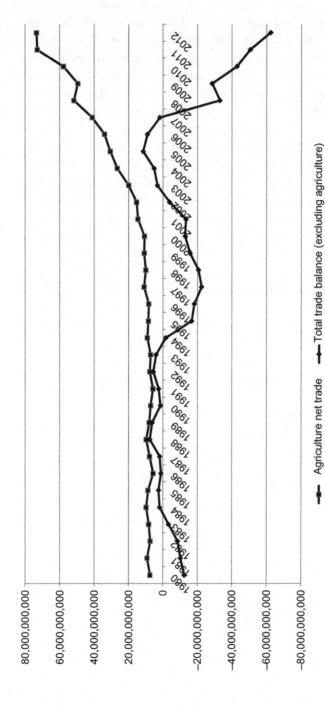

Figure 1.2 Merchandise trade balance, Brazil (US$, 1980–2012). *Source: World Trade Organization (2014).*

but not sufficient for productivity gains to occur, which begs the question of *how* Brazilian farmers overcame many challenges to adopt technology, increase production, and gain international competitiveness. The main contribution of this book is to explain how Brazilian farmers actually achieved this feat. To make productivity gains a reality, Brazilian farmers had to overcome several challenges. For instance, how did they respond to dramatic changes in their economic and competitive environment since the 1990s? What kinds of organizations did they develop to provide missing services and countervail the market power of the middlemen? How did they cope with high transaction costs along agrifood supply chains?[2] How was technology transferred from research labs to the field when public extension services were absent or ineffective? To answer these questions, the book adopts a microanalytic approach and uses case studies and vignettes to *tell the story* of how Brazil became a powerhouse in agriculture. These case studies and vignettes — including the Pinesso story described above — are primarily based on field research conducted in Brazil, complemented with secondary sources of information when available.

The case study analysis of Brazilian farmers and their organizations is organized into three chapters. Each chapter focuses on a different region of the country, with unique characteristics of farming and agrifood system organization and coordination (Figure 1.3). Agriculture in the southern region, which includes the states of Rio Grande do Sul, Santa Catarina,

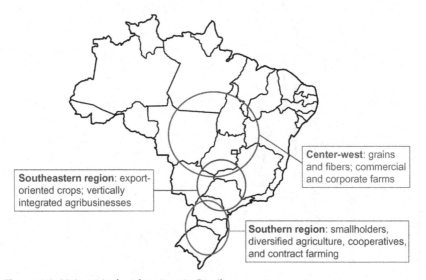

Figure 1.3 Main agricultural regions in Brazil.

and Paraná, has a strong cultural and social influence from European immigrants. It is characterized by traditional family farms with an average size of 40 ha. Farms tend to be diversified with row crops (soybeans, corn, wheat, beans) and livestock (poultry, pork, milk). Another distinguishing characteristic of the southern region is the organization of value chains by cooperatives and contract farming arrangements. The southern states have a strong presence of farmer-owned cooperatives, which provide credit, technical assistance, and farm inputs to their members and collectively market and process their farm production. In addition, poultry and pork producers have contract farming arrangements with processors. Contract farming arrangements were introduced by meat processors as a means to transfer technology to growers, foster production growth, and thereby assure supply. Chapter 3 presents case study evidence about the role of cooperatives and contract farming arrangements in providing missing services, reducing transaction costs, and increasing farm income per hectare, which enabled family farms to increase productivity and remain economically sustainable.

Agriculture in southeastern Brazil is markedly different. Farms tend to be larger, specialized, and vertically integrated. The major crops tend to be export-oriented, such as sugarcane (sugar and ethanol), citrus, coffee, and pulp and paper. The dominant form of organization is the vertically integrated agribusiness enterprise, such as sugarcane mills, orange juice processors, and pulp and paper industries. In addition to vertical integration, these large agribusinesses have contractual relationships with farmers upstream and form complex collaborative ventures downstream in the supply chain. Chapter 4 examines these tightly coordinated agribusiness firms and how they have achieved international competitiveness.

Finally, Chapter 5 describes how the Brazilian center-west region was conquered and became one of the most dynamic agricultural frontier regions in the world.[3] The center-west has a more consolidated farm structure, with an average size of 330 ha. Family farms tend to be large, commercially oriented, and dependent on hired labor. Beginning in the mid-2000s, very large corporate farms — with more than 30,000 ha and featuring specialized, hired labor, professional management, and outside equity capital — started to change the agricultural landscape and increase the level of farm consolidation. Chapter 5 tells the stories of entrepreneurial families who left the southern region in the 1970s and 1980s to colonize the extensive savannahs of Mato Grosso. To overcome common challenges, they formed several for-profit and not-for-profit organizations to adapt technologies to the local conditions, to increase their bargaining power and to lobby for better infrastructure and policies. This chapter explains the role of economies of scale, specialized labor, access to outside capital, and modern farm management technologies in productivity gains achieved by farmers in the Brazilian agricultural frontier.

The last chapter of this book attempts to bring it all together. It summarizes the major findings of the book and concludes with recommendations for policymakers and future research.

NOTES

1. These are quotes from Eugênio from his autobiography, edited by P.N. Souza (2013).
2. An agrifood supply chain includes all stages involved in the production and distribution of a food product, including the pre-farm-gate stages that provide inputs to farmers (such as seeds, fertilizers, crop protectants, machinery, etc.), the actual agricultural production on farms, and the post-farm-gate operations of storage, handling, transportation, processing, and distribution of the final product. An agrifood supply chain involves many participants that must overcome transaction costs and other coordination challenges to provide a quality food product to the end consumer at an affordable price.
3. See, for example, a survey in The Economist (2010).

REFERENCES

Alves, E.R.A., Souza, G.S., Rocha, D.P., Marra, R., 2013. Fatos marcantes da agricultura brasileira. In: Alves, E.R.A., Souza, G.S., Gomes, E.G. (Eds.), Contribuição da Embrapa para o Desenvolvimento da Agricultura no Brasil. Embrapa, Brasília, DF, pp. 15—45.

Bonelli, R., 2013. Notes on productivity and competitiveness. In: Seminar presented at Insper Business School, 01/08/2013.

Companhia Nacional de Abastecimento — CONAB, 2014. Séries Históricas de Área Plantada, Produtividade e Produção, Relativas às Safras 1976—77 a 2014—15 de Grãos. Available from: <www.conab.gov.br> (downloaded 03.10.14.).

Fuglie, K., Wang, S.L., Ball, E., 2012. Productivity Growth in Agriculture: An International Perspective. CAB International, Oxfordshire, UK.

Gardner, B.L., 2002. American Agriculture in the Twentieth Century: How It Flourished and What It Cost. Harvard University Press, Cambridge, MA.

Gasques, J.G., Bastos, E.T., Valdes, C., Bacchi, M.R.P., 2014. Produtividade da agricultura: resultados para o brasil e estados selecionados. Revista de Politica Agricola 23 (3), 87—98.

Organization for Economic Cooperation and Development and Food and Agriculture Organization of the United Nations, 2014. OECD-FAO Agricultural Outlook 2014—2023. Available from: <www.oecd.org/site/oecd-faoagriculturaloutlook/publication.htm> (accessed 05.10.14.).

Rada, N.E., Buccola, S.T., 2012. Agricultural policy and productivity: evidence from Brazilian censuses. Agr. Econ. 43, 355—367.

Souza, P.N., 2013. Eugênio Pinesso: O Caminho de uma Vida. Editora Gibim, Campo Grande, MS.

The Economist, 2010. Brazilian Agriculture, The Miracle of the Cerrado. The Economist Newspaper Limited, 26/08/2010.

US Department of Agriculture, 2014. Production, Supply and Distribution Database. Available from: <apps.fas.usda.gov/psdonline> (downloaded 06.10.14.).

World Trade Organization, 2014. International Trade and Market Access Data. Available from: <www.stat.wto.org> (downloaded 03.10.14.).

CHAPTER 2

Enabling Conditions

Contents

In the early 1970s Brazil was going through a period of rapid economic growth, industrialization, and urbanization. Yet, the agricultural sector in Brazil was still "traditional" with low-intensity production systems, low labor and land productivity, a prevalence of subsistence farms, and rural poverty. In addition, the agricultural sector was taxed by the economic policies of the import substitution model.[1] Agricultural production grew at a modest rate as new land was brought into production but opportunities for agricultural expansion in the traditional areas were limited. Not surprisingly, the country faced constant food supply shortages and was a major importer of agricultural commodities. The country was a recipient of international food aid well into the 1980s.

There was significant room for productivity gains in traditional areas and millions of hectares of arable land yet to be developed in the cerrado with appropriate conditions for farming. With an abundance of natural resources — equivalent to 14.5% of the potential arable land and 13.2% of the renewable water resources in the world — the constraining factors to unlock Brazil's agricultural production potential were human capital, rural credit, and the availability of agricultural technologies adapted to the local weather and soil conditions. Tropical agriculture occurs between latitudes 23°N and 23°S, generally in acid, weathered soils of low fertility.

F. Chaddad: The Economics and Organization of Brazilian Agriculture.
DOI: http://dx.doi.org/10.1016/B978-0-12-801695-4.00002-1

Until Brazilian agricultural researchers and partners developed new crop varieties and agricultural production systems adapted to tropical conditions, it was believed that only temperate regions could feed the world.

This chapter explains the enabling conditions — focusing on natural resource availability, technology, and agricultural policy — that have been commonly associated with the modernization and increasing competitiveness of Brazilian agriculture since the 1970s. We will see how public investments in the 1970s and 1980s provided the necessary conditions for agricultural development in Brazil and how the liberalization of the Brazilian economy in the 1990s and the commodity boom of the 2000s provided strong economic incentives for agricultural entrepreneurs to bring new farmland into production and increase productivity with the use of modern technologies.

2.1 NATURAL RESOURCES

In the beginning, there was a great deal of unproductive land, which included areas that were previously unoccupied or under subsistence or low-input production systems.[2] Brazil has an abundance of natural resources, including 14.5% of the world's equivalent potential arable land[3] and 13.2% of the world's renewable water resources.[4] Weather conditions (rainfall, solar radiation, temperature) are conducive to at least one good crop per year.[5] More recently, with the development of new cultivars and production systems adapted to local conditions, many regions are able to produce two to three crops per year. In most regions, there is no need to irrigate and production can be carried out in rain-fed systems. Soils do not have physical problems as they are deep, with adequate drainage and water retention capacity, and flat, which facilitates mechanization. Differently from other developing countries, Brazil has a low population density outside of the eastern coastal areas.

However, a large share of the country's geographic area — about 200 million hectares — is covered by the cerrado biome, which until the 1970s was considered to be of limited value for agricultural production.[6] Major constraints to agricultural production in the cerrado conditions include acidic soils with low natural fertility and high biological pressure of pests, diseases, and weeds. Cerrado soils are normally highly acidic and nutrient-poor. Decomposition of organic matter in the soil is rapid because of high temperatures and high humidity. Frequent heavy rains cause rapid nutrient leaching and chemical weathering of the soil.

Cerrado soils have deficiencies of a large number of essential plant nutrients, such as P, K, Ca, Mg, and S. Most soils in the cerrado are acidic, with a pH between 4.8 and 5.1. Soil fertility is directly influenced by how acidic the soil is. In areas where soil pH drops below 5, aluminum becomes soluble and accumulates into plant roots, inhibiting root growth and plant life. Another constraint associated with highly acidic soils is phosphorus (P) fixation. This problem is caused primarily by a high content of free ferric oxides, which fix phosphate ions in forms that are unavailable to plants. This soil characteristic causes P deficiency in crops that is difficult to overcome, since added phosphate fertilizers rapidly become fixed in the soil. Approximately two-thirds of all tropical soils in Brazil are too acidic to support traditional food crops and a quarter has high P-fixation (FAO, 2010).

In addition to the low natural fertility of soils, tropical ecosystems like the cerrado are characterized by a high degree of biodiversity and high prevalence of pests and diseases. Winters are mild in the tropics. There is no frost, no snow, and no ice, so there is a high biological pressure of crop pests and parasites year-round. In temperate areas winter eliminates most insect pests prior to the planting of new crops. Crops planted in the spring have a chance to take hold and grow with low biological pressure. In the tropics, crops have to compete with a high population of insects, pests, and weeds. Unless suitable agronomic practices are adopted, monocultures in the tropics are prone to devastation by plant diseases, pests, and natural competition with highly biodiverse ecosystems.

Despite the abundance of natural resources, almost all economies in tropical zones are poor, while those in temperate zones are generally rich. This pattern of temperate zone development and tropical zone underdevelopment is also observed in large countries that cross different ecological zones. For example, Brazil's tropical north, northeast, and center-west regions have historically lower levels of development and income per capita than the south and southeast. Tropical underdevelopment persists because production technology in the tropics has lagged behind temperate zone technology in two critical areas – agriculture and health (refer to Sachs (2001) for details). As a result, productivity is considerably higher in temperate zones than in tropical regions for the major staple crops. Temperate zone economies are food exporters, while tropical zone economies are food importers. Unfortunately, most temperate zone agricultural technologies are inappropriate for tropical areas because they are ecologically specific and do not diffuse easily across ecological zones.

2.2 TECHNOLOGY

Brazil was the first country to invest heavily in agricultural technologies and production systems adapted to tropical conditions.[7] The first challenge was to identify and quantify the limiting factors to agricultural development in the cerrado. Subsequently, strategies and technologies adapted to local conditions had to be developed and disseminated to overcome these challenges. The development of the Brazilian cerrado into productive agricultural land required a portfolio of technologies, including new plant varieties and hybrids, soil fertility improvements, strategies to control diseases, pests, and weeds, use of no-tillage systems, and integrated crop and livestock systems. These technologies, developed over the last 40 years, removed the constraints to producing high-yield crops in the poor, acid soils of the cerrado. If it were not for the technological breakthroughs that made the soils of the cerrado more productive, Brazil would not have achieved the productivity gains documented in Chapter 1.

2.2.1 Plant Breeding and Genetics

Work on genetic resources and traditional plant breeding enabled the adaptation of crops to the unique soil and ecological conditions of the cerrado. The research efforts of Brazilian scientists made it possible to develop crop varieties adapted to lower latitudes, which are capable of producing yields as high as those produced in temperate regions. By the late 1990s, the yield potential of seeds developed by Brazilian researchers was far higher than the actual productivity achieved by producers. Through plant breeding, Brazilian researchers have been able to "tropicalize" several crop systems from corn and soybeans to apples and grapes. A recent breakthrough was the development of precocious varieties of soybeans and corn which allow for double cropping in the same field.

These plant breeding efforts were led by pioneers such as Dr. Romeu Kiihl, who is considered the "father of soybeans" in Brazil. Dr. Kiihl received his bachelor's degree in agronomy from the College of Agriculture at the University of São Paulo (ESALQ/USP) in 1965 and then moved to the United States to pursue his graduate studies at Mississippi State University in plant breeding and genetics. There he worked with another pioneer, Dr. Edgar Hartwig, who developed soybean varieties adapted to the southern United States. Upon his return to Brazil in the early 1970s, Dr. Kiihl started to develop soybean varieties to be used as a winter crop in rotation with rice and other summer crops, first at the Campinas Agronomic Institute (IAC) and then at the Paraná Agronomic Institute (IAPAR), two well-respected state research institutes. The soybean cultivars locally bred by Brazilian researchers in the 1970s and 1980s from cultivars imported from the United States allowed soybean to become a major crop in the southern region.

In 1978 Dr. Kiihl was hired by the Brazilian Agricultural Research Corporation (EMBRAPA) soybean center in Londrina, Paraná, to lead the research efforts to develop soybean varieties adapted to the lower latitudes of the cerrado. This required an effort to identify genes that expressed late flowering under short day conditions − known as the long juvenile trait. New cultivars were bred with a range of sensitivity to day length, which allowed soybean to adapt to diverse growing conditions and to move northward to the cerrado. While EMBRAPA coordinated these basic research efforts, the development of commercial seeds adapted to local conditions was carried out by state research institutes across the country and producer-owned organizations such as Coodetec in Paraná and Fundação Mato Grosso (FMT). This research network developed new varieties that were resistant to local disease and pest problems and provided additional genetic diversity, thereby allowing genetic improvements with greater yield potential. We will see later in the book how Coodetec and FMT played an important role in developing new cultivars and making them available to farmers.

More recently, Dr. Kiihl became the scientific director at Tropical Melhoramento e Genética (TMG), the commercial seed company controlled by FMT. Dr. Kiihl's professional trajectory illustrates the breadth of research institutions in Brazil − both public and private − that were responsible for developing the technologies that allowed the country to become an agricultural powerhouse. It is also important to note that efforts to develop agricultural technologies in Brazil preceded the foundation of EMBRAPA in 1974 and that many Brazilian researchers received training in domestic and international universities.

Until the mid-1990s R&D efforts in plant genetics and breeding were mainly carried out by public institutions as Brazil did not have suitable protection for intellectual property rights (IPRs). These public research institutions, in turn, had agreements with research foundations and cooperatives maintained by producers throughout the country, including Coodetec in Paraná and FMT in Mato Grosso. These producer-controlled organizations conducted field trials to test new seeds coming from the research labs and identify seed varieties that adapted best to local conditions, which were then multiplied and distributed to producers.

Because Brazil signed the 1978 Act of the International Convention for the Protection of New Varieties of Plants (UPOV), producers could save seeds for later use without having to compensate the plant breeder's rights.

It was only after Brazil signed the WTO agreement on Trade Related Aspects of Intellectual Property Rights (TRIPS) in 1994 that it started to implement laws and regulations to effectively protect private IPRs. The main laws that constitute the domestic institutional framework for the protection of plant breeders' rights include the Intellectual Property Law of 1996, the Plant Variety Protection Law of 1997, the Seed Law of 2003 that created the National System of Seeds and Seedlings, and the Biosafety Law of 2005 that regulates research, production, and marketing of genetically modified organisms (GMOs).[8] This set of laws provided reasonable protection for private investments in agricultural R&D, allowing private firms to appropriate some of the economic benefits of innovation with royalties and technological fees.

With the growth of the agricultural sector, this institutional framework fostered the entry of new players in the domestic seed market. Multinational companies acquired domestic seed companies and entered licensing agreements with public institutions, such as EMBRAPA, that have access to the local germplasm.[9] As a result, the number of protected plant varieties increased from 51 in 1998 to 1,708 in 2012.[10] The domestic seed market increased from 1.6 million tons in 2001 to more than 3 million tons in 2013. Despite this growth, the traditional industry structure based on public research institutions and producer-owned organizations was disrupted by new entrants with more focused research programs and more aggressive commercial teams. EMBRAPA and other public research institutes still play a relevant role in the development of new seed varieties but the market structure has changed considerably since the late 1990s. The market share of private companies, and of multinational firms in particular, has increased substantially, especially for crops that use GM traits such as soybeans, corn, and cotton.[11] Monsanto, for example, launched in 2013 the first GM soybean seed developed uniquely for the Brazilian market with stacked genes for herbicide tolerance and protection against major worms. The development of this product required 11 years of research, field trials, and regulatory approvals.

2.2.2 Soil Fertility and Conservation

Brazilian researchers have learned how to use the tropical soils of the cerrado. Aluminum toxicity is countered by adding lime and gypsum to the soil, which neutralizes soil acidity and renders the aluminum inert. Slow-release forms of phosphorus were developed, such as the addition of rock phosphate, to ameliorate the problem of phosphorus fixation. With application of the correct amounts of lime, phosphorus, and other nutrients, soil fertility could be improved and maintained over time, leading to crop yields comparable to those in temperate zone countries. In 1979, the Japan International Cooperation Agency (JICA) launched a US$300 million program to help transform Brazil's cerrado into fertile agricultural lands with the use of lime and gypsum to counter the soil's acidity. These efforts resulted in taking land with very low levels of productivity and making them effectively productive.

In addition to improving soil fertility, new technologies were developed to conserve land resources. One example is minimum tillage farming, which is a system of planting crops into untilled soil by opening a narrow slot, trench, or band of sufficient width and depth to obtain proper seed coverage. No other soil tillage is done, which lessens soil erosion — a major agronomic problem of tropical agriculture. In addition, conservation farming using zero tillage is associated with 75% less fuel and 30% less water use in agriculture, as well as with carbon sequestration in the soil. The area under no-tillage in Brazil increased from 1.4 million hectares in the late 1980s to 25.5 million hectares in 2007, which places Brazil as the number-one country in area under no-tillage farming, ahead of the United States, Argentina, and Canada (FAO, 2008). Conservation farming is another example of a knowledge-intensive and location-specific technology that required investments in research on suitable varieties, management practices adapted to specific sites, appropriate machinery, and extension services.

2.2.3 Biological Nitrogen Fixation (BNF)

Nitrogen is the main building block of protein, such as muscle in mammals and plant tissue in crops. If the level of nitrogen in the soil is increased, plant growth can be significantly enhanced. Legumes are a group of plants that interact with bacteria (rhizobia) in the soil to fix nitrogen from the air, deposit it into the soil, and make it available for other plants to use. The nitrogen deposited by legumes can be readily converted into larger harvests and reduce chemical nitrogen fertilizer application. Soybean varieties that require no nitrogen have been selected for use in Brazil with yields of up to 6 tons per hectare. This is done with the inoculation of soybean seeds with *Bradyrhizobium*, a species of bacteria that is able to fix atmospheric nitrogen into forms readily available for plants to use. Intercropping with a legume, such as soybeans, has the potential to decrease the need for chemical fertilizer application. Brazilian researchers were also able to develop biological nitrogen fixation (BNF) techniques for non-legume plants that do not normally fixate nitrogen, such as tropical pastures, rice, and sugarcane, with positive results in replacing part of the requirements for nitrogen fertilizers with BNF.

2.2.4 Crop Rotation and Integrated Pest Management

Crop rotation is the cornerstone of pest control in the tropics. When a single crop is planted repeatedly in the same soil, insects and diseases that attack that crop are allowed to build up to unmanageable levels, greatly reducing the farmer's harvest. The most basic form of crop rotation is the following: never plant the same crop in the same place twice. This results in naturally breaking the cycles of weeds, insects, and diseases that attack food crops. Rotations are used to prevent or at least partially control several pests and at the same time to reduce the farmer's reliance on chemical pesticides. Crop rotation is often the only economically feasible method for reducing insect and disease damage in the tropics. Crop rotation replaces a crop that is susceptible to a serious pest with another crop that is not susceptible. Each food crop comes with its own set of pests that attack that particular crop. By planting a different crop each time, the farmer is able to starve out those pests. Often a set of three or four crops are planted on a rotating basis, ensuring that by the time the first crop is replanted, the pests that attack it are substantially reduced. Another side benefit of crop rotation is that it improves the soil. Constantly growing the same crop in the same location will strip the soil of the nutrients that a particular crop requires. Rotating to a different crop also reduces the demand on soil fertility.

2.2.5 Integrated Crop–Livestock Systems

Integrated crop–livestock systems, whereby crop residues feed livestock and livestock manure fertilizes crops, are an example of resource-saving technology with several agronomic and environmental benefits, which include improved soil properties and fertility, reduced incidence of diseases, pests, and weeds, improved nutritional value of forage, and higher crop and animal productivity. Brazilian researchers are now extending integrated farming systems to forestry, so that crop cultivation and livestock farming can occur in forested areas.

2.2.6 The Development of the Cerrado

Following these agricultural technology developments, the cerrado was opened up to agriculture and new land was brought into production. Between 1975 and 1996, the area in planted pastures increased 4.8% on average each year, from 17.8 to 49.2 million hectares. During the same period, the area used in row crops increased from 9.2 to 13.1 million hectares, equivalent to an average annual growth rate of 1.7%. About half of the cerrado native vegetation had been converted to agricultural land by 2005 (Table 2.1).[12] The development of the cerrado into new agricultural land was responsible for the largest share in food production between 1970 and 1990. Since then, most of the production increases in the cerrado were due to yield growth (Table 2.2). Grain production in the cerrado increased from 8 million tons in 1970 (equivalent to 35% of the total Brazilian production) to 48.2 million tons in 2006 (or 49% of

Table 2.1 Land use in the cerrado

Land use	Area (million hectares)	Share of total area (%)
Native/undisturbed	116.1	56.9
Planted pastures	65.9	32.3
Cropland	18.0	8.8
Urban land	3.0	1.5
Other uses	1.0	0.5
Total	204	100

Source: Lal (2008).

Table 2.2 Evolution of average yields for major crops in the Brazilian cerrado (1975–2005)

	Average yield (ton/ha)			Yield growth (1975–2005) (%)	Potential yield (ton/ha)[a]
	1975	1993	2005		
Soybean	1.32	2.20	2.81	113	4.5
Maize	1.57	2.70	4.36	177	10.5
Upland rice	1.03	1.20	2.32	125	3.5
Phaseolus beans	0.48	0.71	1.83	281	3.0
Irrigated wheat	2.80	3.95	5.23	87	7.0
Cotton	1.60	2.63	3.64	127	4.5
Coffee	0.82	1.33	2.35	187	3.5

[a]Average research fields.
Source: Spehar (2008).

the total). In 2006, the cerrado produced 89% of the cotton, 69% of the sorghum, 55% of the cattle, 53% of the soybeans, 48% of the coffee, 37% of the rice, and 30% of the corn produced in Brazil.[13]

2.3 AGRICULTURAL POLICY

Agricultural policy changes implemented in the late 1960s unlocked the production potential of the cerrado. Most of the technologies documented above were developed in Brazilian research labs and fields, under the coordination of EMBRAPA. Agricultural policy also provided the necessary conditions for farmers to adopt these technologies and develop the cerrado, including rural credit, agricultural price support, and extension services.

Agricultural policy goals and programs in Brazil have changed significantly in the last 50 years (Table 2.3). The period between the mid-1960s and the early 1980s was characterized by massive government intervention in agriculture, primarily by means of subsidized rural credit and price support mechanisms including government purchases and storage of excess supply (Figure 2.1). At that time, the agricultural sector in Brazil was in general not competitive — except in tropical products such as coffee and sugar — and characterized by highly skewed distribution of farm income and land ownership. Production growth was mainly achieved with continuous incorporation of agricultural land. It was in the 1960s and 1970s that the country started to urbanize, as many rural poor migrated to large cities. During this period, agricultural policy had the objective of promoting the food security of an increasingly urban population while compensating the agricultural sector for the anti-export bias of the import substitution model.

It was at that time that policymakers decided to promote the modernization of Brazilian agriculture based on three pillars: investments in agricultural research and extension, subsidized rural credit, and price controls. The Brazilian Agricultural Research Corporation (EMBRAPA) was created in 1974 as a public enterprise to coordinate a public network of state-level research institutes and universities. EMBRAPA was designed as the "research arm" of the Ministry of Agriculture, Livestock, and Food Supply (MAPA) and received the lion's share of Brazilian public investments in agricultural research and development. The public investments in agricultural R&D through EMBRAPA increased substantially in the 1970s and then peaked in 1996 at BRL 1.4 billion (Figure 2.2). In its

Table 2.3 Evolution of agricultural policy in Brazil

	1965–1985	1985–1995	1995–2010
Macroeconomic conditions and policy	• High inflation • High growth rate • Controlled exchange rate • Import substitution model	• Uncontrolled inflation • Low growth • Debt crisis • Heterodox plans	• Control of inflation • Modest growth rate • Volatile exchange rate • High real interest rates • Privatization
Agricultural policy goals	• Food security • Increased government expenditure in farm policy	• Deregulation • Liberalization • Decreased government expenditure in farm policy	• Land reform programs • Family farming and social inclusion • Relatively low agricultural support
Price support and government storage	• Massive intervention: public agencies, government purchases and storage, price controls • Commodity price support (PGPM)	• Decreased intervention • Agricultural commodity market deregulation	• Modest and selective intervention
Rural credit	• Government supply of credit financed by Treasury (SNCR) • Negative real interest rates	• Decreased government supply of credit • Interest rates less subsidized	• Credit lines targeted to family farms (PRONAF) • Specific programs for investment credit (BNDES) • Agricultural credit crisis and debt rescheduling

(Continued)

Table 2.3 (Continued)

	1965–1985	1985–1995	1995–2010
Agricultural trade policy	• Closed economy • High tariffs • Export taxes on primary commodities	• Unilateral openness to trade • Regional integration (Mercosur) • Elimination of export taxes	• Aggressive policy against agricultural trade barriers • WTO dispute panels • Leadership in G-20 and focus on south–south trade
Agricultural research and extension	• High investments in public research (EMBRAPA) • Development of public extension service network (EMBRATER)	• Leveling-off of public investment in public R&D • Extinction of EMBRATER	• Crisis of public research and extension services • Renewed commitment to public R&D investments • Extension targeted to family farms
Social policies (family farms and land reform)	• Minimal	• Initial stage (Extraordinary Ministry of Land Reform)	• Ministry of Agrarian Development (MDA) • Distributive programs: land reform, PRONAF, *Bolsa Família*, rural retirement

Source: Adapted from Chaddad and Jank (2006).

formative years, EMBRAPA also received funding from international institutions such as the World Bank, the Inter-American Development Bank, and the US Agency for International Development. In the process of developing and expanding its programs, EMBRAPA was sending as many as 1,000 of its staff for graduate training each year — within Brazil and abroad. Since the mid-1990s, public funding for agricultural R&D has decreased in real terms to about 1% of the sectoral gross product. This is substantially lower than in developed countries that invest 2–3%

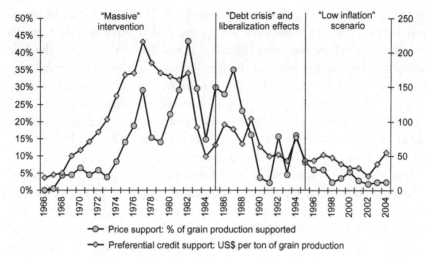

Figure 2.1 Commodity price and preferential credit support to farmers in Brazil (1966–2004). *Source: Chaddad and Jank (2006) based on data from the Ministry of Agriculture, Livestock, and Food Supply (MAPA).*

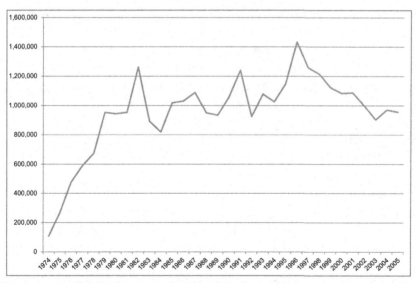

Figure 2.2 Total EMBRAPA expenditures in BRL thousands (1974–2005) (real values corrected for inflation with IGP-DI; values expressed in 2005 BRL). *Source: Author's elaboration using data available in Gasques et al. (2006).*

of the gross value of agricultural production in R&D. Although its financial support has declined over the years, EMBRAPA is still a robust organization. Many observers believe that EMBRAPA is one of the best agricultural research systems in the developing world and a leader in tropical agricultural research.[14] Brazilian researchers were responsible for 6% of the world's scientific production in agricultural and biological sciences in 2013, up from 1.5% in 1996, which is a much higher share than in other fields of knowledge (according to the The SCImago Journal and Country Rank (http://www.scimagojr.com/)).

As the sister organization to EMBRAPA, the Brazilian Technical Assistance and Rural Extension Corporation (EMBRATER) was set up in 1975 to provide agricultural extension services throughout the country. Like EMBRAPA, EMBRATER was designed to coordinate a national system of technical assistance and rural extension with 26 state-level organizations, but it was extinguished in 1990. Between 1990 and 2003, the federal government played no role in disseminating agricultural technologies to farmers, which was the sole responsibility of poorly funded state-level organizations. With the crisis of the public agricultural extension system, producers increasingly depended on farmer-owned cooperatives, farm input retailers, and food processors to have access to information, technology, and technical assistance. It was only in 2003 that the federal government redesigned a national agricultural extension policy focused on family farmers.

The National Rural Credit System (SNCR) was established to provide rural credit at below-market rates to farmers. Initially, agricultural credit was mainly provided by the federal government through public banks, such as *Banco do Brasil* and *Banco do Nordeste*. Interest rates were fixed to protect farmers from high inflation, but the system created the need for direct subsidies from the National Treasury to remain financially viable. Between 1975 and 1984, preferential credit support to farmers was above US$150 per ton of grain produced in the country (Figure 2.1). The availability of agricultural credit peaked at BRL 123 billion in 1979 and then had to be reduced as part of the macroeconomic adjustments of the 1980s and 1990s, reaching the lowest level in 1996 at BRL 18 billion. Since then the availability of agricultural credit increased again and reached BRL 87 billion in 2010 (Figure 2.3). In 2012, the rural credit system (SNCR) was comprised of 298 federal, state, and cooperative banks providing government-supported credit to agriculture. The system was controlled, coordinated, and supervised by the Brazilian Central Bank.

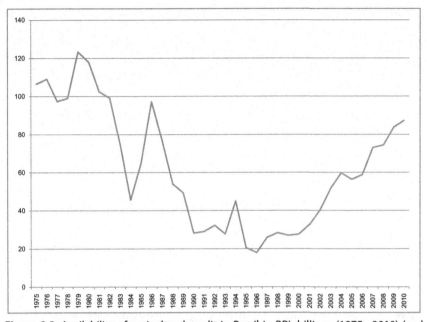

Figure 2.3 Availability of agricultural credit in Brazil in BRL billions (1975–2010) (real values corrected for inflation with IGP-DI; values expressed in 2010 BRL). *Source: Author's elaboration using data collected from the Statistical Yearbook of Rural Credit, various years, Brazilian Central Bank (BACEN).*

In addition to subsidized rural credit, the Guaranteed Minimum Price Policy (PGPM) established floor prices for many agricultural commodities to compensate farmers against the biases of the import substitution model. The minimum price policy was based on government acquisition and storage of agricultural products, which was the responsibility of the National Food Supply Company (CONAB). Between 1975 and 1984, at the height of this massive intervention policy model, CONAB purchased 20–45% of the grain produced in the country, leading to huge and costly public stocks (Figure 2.1). More recent data suggest that the minimum price policy (PGPM) based on CONAB purchases of excess commodity supply has been significantly reduced (Figure 2.4).

The debt crisis of the late 1980s forced the Brazilian government to decrease support to farmers and to review agricultural policy goals. Economy-wide structural reforms introduced in the early 1990s further decreased the distortions of agricultural policy in Brazil by eliminating export taxes and price controls, deregulating and liberalizing commodity markets, unilaterally reducing trade barriers, and introducing private instruments for agricultural

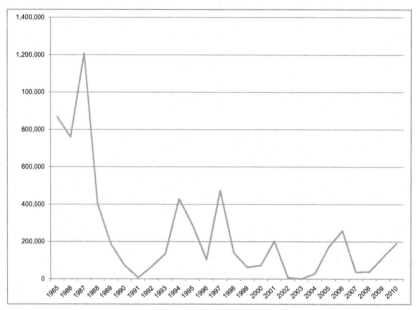

Figure 2.4 Federal government purchases of agricultural commodities in metric tons (1985–2010). *Source: Author's elaboration using data collected from the CONAB website at http://www.conab.gov.br/.*

financing. Domestic agricultural prices declined below international levels in the 1990s and price support to farmers between 1995 and 1999 was negative (OECD, 2005). Low commodity prices, coupled with high farm indebtedness, led to a massive rural debt crisis in 1995 and again in 2005.

Following these policy changes, government expenditure on agriculture-related programs in Brazil decreased relative to both total government expenditure and GDP (Figure 2.5). The annual average amount spent in agricultural policy programs, in inflation-adjusted values, reached BRL 20.8 billion in the late 1980s, which represented 5.6% of total government expenditure (Table 2.4). The average amount spent on agricultural programs decreased to BRL 10.3 billion per year, representing 1.8% of total government expenditure in 2000–2004. Despite recent increases in government expenditure to an annual average of BRL 14.6 billion between 2005 and 2009, agricultural policy continued to lose relevance relative to total government expenditure.

Not only has government expenditure on farm policy decreased by half in real terms since the late 1980s, but agricultural policy goals were broadened after 1995. In addition to increasing agricultural production and

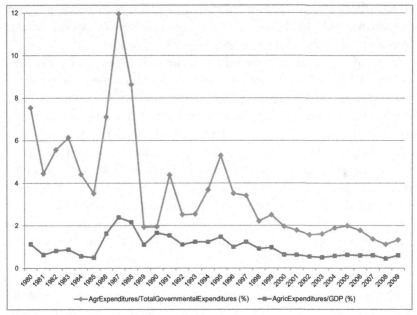

Figure 2.5 Government expenditures on farm support as a share of total government expenditures and GDP (1980–2009). *Source: Author's elaboration using data available in Gasques et al. (2006, 2010).*

Table 2.4 Brazilian government expenditures in farm programs by function (1980–2009)[a]

	Average annual expenditure (BRL million)			Expenditure shares (%)		
	Agriculture (A)	Agrarian organization (B)	Total (C)	Agriculture/ total (A/C)	Agrarian organization/ total (B/C)	(A + B)/Total government expenditure (A + B/C)
1980–1984	8,790	131	8,921	98.5	1.5	5.7
1985–1989	19,549	1,330	20,879	93.6	6.4	5.6
1990–1994	17,510	1,229	18,739	93.4	6.6	2.8
1995–1999	15,042	3,088	18,130	83.0	17.0	3.2
2000–2004	8,262	2,095	10,357	79.8	20.2	1.8
2005–2009	10,343	4,268	14,611	70.8	29.2	1.5

[a]Expenditures are measured in BRL million corrected for inflation by IGP-DI (base year is 2005).
Source: Author's calculations based on data provided in Gasques et al. (2006, 2010).

productivity, policy goals included land reform and family farming in an attempt to alleviate rural poverty. This shift in agricultural policy goals led to increasing government expenditure in a new focus area called agrarian organization.[15] Concomitant to budget reductions in traditional agricultural policies, federal government expenditure on agrarian organization programs increased to BRL 4.2 billion in 2009, representing almost 30% of total expenditure on farm programs in 2005–2009 (Table 2.4).

Agrarian organization programs are primarily related to land reform. Between 1995 and 2010 approximately 1.2 million new family farms were settled in expropriated land across the country. In addition to land reform, the government adopted a set of policies targeted at family agriculture in 1995 — known as PRONAF — including subsidized credit lines, capacity building, and research and extension services. Interestingly, the Brazilian government created a new ministry in 2000 to run programs targeted at family farms and land reform — the Ministry of Agrarian Development (MDA). Brazil is probably the only country in the world with two ministries of agriculture. This reflects a supposed duality of farming in the country and the misleading perception that agribusiness development necessarily leads to small farmer exclusion.

The policy goal shift toward agrarian organization was associated with significant changes in resource allocation. First, traditional agricultural policy functions were sacrificed to support agrarian organization programs. Table 2.5 shows OECD estimates for the general services support (GSSE) received by Brazilian farmers, which includes investments in research and development (R&D), extension, infrastructure, and other public goods to the benefit of all farmers. The data show a significant, inflation-adjusted reduction of the GSSE from BRL 6.8 billion in 1995–1999 to BRL 2.6 billion in 2010–2012. Government expenditure on land reform reached BRL 2.6 billion in 1995–1999 or 39% of total GSSE. Although land reform expenditure was reduced to BRL 1.2 billion in 2010–2012, the program still represented 44% of GSSE. At the same time, expenditure on other general services to farmers suffered significant budget cuts since 1995, including R&D, agricultural education, inspection services, and other infrastructure investments not related to land reform such as irrigation, electrification, and rural housing.

Second, government expenditure was dispersed in an increasing number of programs. The number of agriculture-related programs increased from 30 before the year 2000 to 100 programs in 2003, 84 under the function "agriculture" and 16 under the function "agrarian organization." The performance of many of these programs was not subject to cost–benefit evaluation, expenses were quite variable or even arbitrary, and many programs were stretched to the limit and could not contribute to intended goals with continued budget reductions.

Table 2.5 Government support to agriculture in Brazil in BRL millions (1995−2012)[a]

	1995−1999	2000−2004	2005−2009	2010−2012
Total support estimate (TSE)	−6,873	12,469	15,070	14,389
Producer support estimate (PSE)	−13,815	8,916	11,155	11,310
General services support estimate (GSSE)	6,886	3,475	3,508	2,691
Research and development	1,182	921	343	336
Agricultural schools	478	511	532	351
Inspection services	268	155	154	189
Infrastructure (not including land reform)	1,466	544	207	75
Land reform and settlement	2,692	1,141	1,867	1,175
Marketing and promotion	40	18	89	212
Public stockholding	759	163	304	353
Miscellaneous	0	21	12	0

[a]Expenditures are measured in BRL millions corrected for inflation by IGP-DI (base year is 2005).
Source: Author's calculations based on data available in OECD (2014).

As a result of these policy changes, government support to agriculture has been relatively low in Brazil since 1995 in both absolute and relative terms. Agricultural support measured by the producer support estimate (PSE)[16] has increased since the 1990s and averaged US$8.7 billion between 2010 and 2012, but it has remained at about 4.5−5.5% of total farm receipts since 2000 (Table 2.6). The combined PSE for developed countries totaled US$257 billion and represented 18.8% of total farm receipts in 2010−2012. Developing countries such as China, Indonesia, and Russia also have considerably increased support to agriculture since 2000 and now have producer support as a share of gross farm income comparable to rich economies.

Another measure of support to agriculture is the total support estimate (TSE), which is the monetary value of all gross transfers from taxpayers and consumers related to policy measures that support agriculture. In addition

Table 2.6 Producer support estimates in US$ millions in selected countries (1995–2012)

	PSE (USD million)[a]				PSE (%)[b]			
	1995–1999	2000–2004	2005–2009	2010–2012	1995–1999	2000–2004	2005–2009	2010–2012
Australia	1,109	865	1,372	1,297	5.2	3.8	4.1	2.7
Canada	3,507	4,842	6,352	7,462	16.5	20.1	17.9	15.3
Chile	464	360	310	346	9.0	7.5	3.9	2.6
EU-27	114,430	101,800	126,284	107,960	35.0	32.8	26.0	19.4
Japan	54,854	47,409	41,561	60,873	58.5	57.3	49.8	53.7
Korea	19,936	17,278	20,308	19,482	64.7	60.3	54.2	49.3
Mexico	3,045	6,773	5,787	6,800	10.1	19.8	12.9	12.5
New Zealand	55	40	86	134	0.9	0.5	0.8	0.8
United States	36,415	44,629	33,659	30,875	16.8	19.0	11.2	7.8
OECD	**257,245**	**243,566**	**254,669**	**256,916**	**31.4**	**30.0**	**23.2**	**18.8**
Brazil	−5,115	2,523	5,920	8,728	−9.2	4.9	5.5	4.6
China	1,998	18,163	50,186	135,386	0.9	5.8	8.7	15.0
Indonesia	−2,182	2,684	2,996	23,570	−16.4	8.8	5.8	18.8
Kazakhstan	404	200	873	1,565	12.1	5.9	10.2	11.6
Russia	4,605	3,733	11,975	14,979	12.3	10.6	17.1	16.7
South Africa	901	592	765	499	10.4	6.9	5.6	2.5
Ukraine	−560	217	1,662	280	−3.9	1.8	7.2	1.2
Other economies	**52**	**28,113**	**74,376**	**185,007**	–	–	–	–
Total	**257,297**	**271,679**	**329,045**	**441,923**	–	–	–	–

[a]Agricultural support is defined as the annual monetary value of gross transfers to agriculture from consumers and taxpayers, arising from policies that support agriculture. Transfers included in the producer support estimate (PSE) are composed of market price support, budgetary payments, and the cost of revenue forgone by the government and other economic agents.
[b]The percentage producer support estimate (PSE %) represents policy transfers to agricultural producers, measured at the farm gate and expressed as a share of gross farm receipts.
Source: Author's calculations based on data available in OECD (2014).

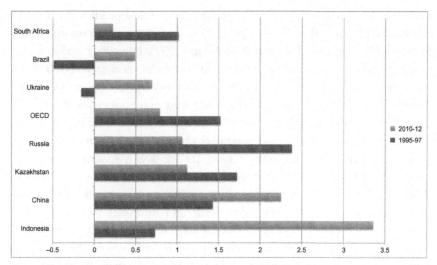

Figure 2.6 Total support estimate as a share of GDP in selected emerging economies and OECD average (1995–1997 and 2010–2012). *Source: Author's calculations
based on data available in OECD (2014).*

to the PSE, the TSE also includes expenditure in general services to the
collective benefit of farmers such as R&D, training, and infrastructure
(GSSE), and transfers to poor consumers in the form of food subsidies.
The TSE for Brazilian agriculture was 0.49% of GDP in 2010–2012,
much lower than the rich country average and in other emerging economies (Figure 2.6). These numbers suggest that Brazil is among the countries with the lowest levels of support to agriculture. Most of the
government support goes to producers in the form of price and income
support, preferential credit, and social programs such as PRONAF and land
reform. A declining share of the total support is dedicated to general services to the sector as a whole.

2.4 SUMMARY

In retrospect, farm policies in Brazil have evolved in the last three decades
from a food security and self-sufficiency emphasis until the late 1980s, to
deregulation and trade liberalization in the early 1990s and, since then,
placed a higher priority on social policies targeting small family farms and
land reform. Agriculture benefited from the change in development paradigm, openness to trade, and macroeconomic stabilization with production
growth, productivity gains, and modernization. Brazilian farmers had to

become increasingly efficient to be able to survive and prosper in an environment with low subsidies, low protection, and increasing economic integration. The rural debt crises of 1995 and 2005 effectively weeded out many producers that were highly leveraged. Producers that survived these crises were able to benefit from the commodity cycle of the 2000s and prosper. Not surprisingly, Brazilian farmers were able to increase the rate of agricultural productivity gains in the last decade as shown in Chapter 1.

Despite these recent gains, several challenges remain. Production and productivity growth in Brazilian agriculture are largely explained by technological change,[17] but this process has not been even across the country. There has been some convergence in productivity of crops like corn, rice, and wheat across regions, but the north and northeast regions still lag behind the national averages. According to the latest Census of Agriculture, about 440,000 farm establishments (or 8.4% of the total) are responsible for 85% of the total production value. As many as 3.2 million farmers − 63% of the total number of establishments − produce less than 4% of the total production value and live in extreme poverty (IBGE, 2006). This structural heterogeneity between farm establishments is largely explained by the absorptive capacity of farmers. Those who adopted modern technologies and production systems achieved high productivity growth and thus were able to grow and prosper. The majority of farmers, however, did not have access to formal education and technical assistance or were less entrepreneurial and thus lagged behind.[18] The declining public investments in general services to farmers − such as rural education, extension, and infrastructure − will only exacerbate this problem.

In addition, access to capital continues to be a problem for many farmers as a result of high real interest rates. This problem has been partially addressed with the recent increases in the availability of subsidized agricultural credit. However, market failures still remain that make it difficult for many farmers to access credit.[19] Poor infrastructure is a chronic problem in rural areas, particularly in the frontier regions of the cerrado. Recent decreases in general support to farmers in the form of agricultural R&D, extension services, education, infrastructure, and inspection services might dampen future production and productivity growth across all regions and all types of farms.

As agricultural policy evolved and producers increased production and productivity, the agrifood sector in Brazil underwent significant structural changes. First, it was exposed to a dramatic "competition shock" as a result of economic liberalization, industry deregulation, and dismantling of the safety net provided by massive government expenditure in traditional agricultural policy programs. Subsequently, it experienced significant modernization and industrialization induced by private sector strategic responses to these institutional and policy changes. The development of a global agrifood model in Brazil resulted in structural changes in all stages of the agrifood value chain, significant export-led growth, and apparent small farmer exclusion. In what follows, this book describes and analyzes in greater detail how alternative agrifood chains were organized in different regions, adapted to these changes and achieved dynamic international competitiveness based on technological and organizational innovations.

NOTES

1. Starting in the 1950s, Brazil epitomized the use of import-substituting industrialization policies to promote economic growth and development. Such policies included high levels of protection and subsidies for the manufacturing sector, taxes on agricultural exports, and an over-valued currency, which is in itself an implicit export tax.
2. Under the low-input system, production is based on the use of traditional cultivars, labor-intensive techniques, no application of fertilizers, and no use of chemicals for pest and disease control. This traditional farming system is largely subsistence-based and not necessarily market-oriented. Until the 1980s, farming in the cerrado was based on extensive livestock production and subsistence production of staple crops such as rice, beans, and cassava.
3. The most fundamental factor influencing the agricultural production capacity of a country or region is its potential arable land, which is responsible for growth of the major food crops. In 2000, the Food and Agriculture Organization of the United Nations estimated the potential arable land for rain-fed agricultural production in 160 countries (FAO, 2010). The gross potential arable land in Brazil was estimated to be 549 million hectares, which is equivalent to 13.2% of the world's total (4,144 million hectares). The gross potential arable land was adjusted for two non-agricultural uses — i.e., protected land and land for human settlement — to give the net values for potential arable land. The net potential arable land was then adjusted for "quality" — i.e., suitability for agricultural production. The quality-adjusted arable land is the equivalent potential arable land. Brazil has 394 million hectares of equivalent potential arable land or 14.5% of the world's total (2,718 million hectares). A more recent study commissioned by the FAO shows that Brazil has 176 million hectares of suitable non-cropped, non-protected land available for agricultural expansion (or 14.4% of the world's total), of which 45 million hectares are non-forested (Fischer and Shah, 2010).
4. The FAO estimates that Brazil has a volume of 5,661 km^3 of internal renewable freshwater resources, which is equivalent to 13.2% of the world's total (FAO, 2014).
5. Most of the Brazilian territory lies in the tropical zone to the north of the Tropic of Capricorn (latitude 23°S), with monthly mean temperatures above 18°C. The southern states of Paraná, Santa Catarina, and Rio Grande do Sul have subtropical weather conditions, with one or more months with mean temperatures below 18°C but above 5°C. Most of the country faces no climate constraints to agricultural production (Fischer et al. 2002).

6. The cerrado is a savannah-like vegetation of low trees, scrub brush, and grasses. It occurs entirely within Brazil and covers 204 million hectares or 23% of Brazil's land area. The cerrado is the second largest biome in Brazil, after the Amazon biome, which covers 350 million hectares. The cerrado is the tropical savannah with the richest biodiversity in the world, as it comprises one-third of the national biodiversity and 5% of the world's flora and fauna. The cerrado biome, by virtue of climate and soils, is more suited for intensive grain cropping and livestock production than the tropical rainforest biome (Faleiro and Farias Neto, 2008).

7. Public investments in agricultural research and development are discussed in the following section on agricultural policy.

8. Although the first genetically modified (GM) crop was approved by the US Food and Drug Administration (FDA) in 1994, and the first herbicide-resistant soybean seed was introduced in the US market in 1996, it was only 10 years later that Brazil implemented regulations regarding GMOs. In 2014, Brazil had 40 million hectares of GM crops, second only to the US with 70 million hectares.

9. Germplasm is any living tissue from which new plants can be grown, including a seed or another plant part (e.g., a leaf, a piece of stem, or pollen) that can be turned into a whole plant. Germplasm contains the genetic information of a particular plant species, a valuable natural resource of plant diversity. The genetic diversity of germplasm gives plant breeders the ability to develop new varieties that are more productive or can resist constantly evolving pests, diseases, and environmental stresses. EMBRAPA has the largest germplasm in Brazil, with more than 100,000 specimens of 600 different species with food and agricultural importance.

10. The Ministry of Agriculture maintains a database with the national registrar of plant varieties available at http://www.agricultura.gov.br/vegetal/registros-autorizacoes/protecao-cultivares/cultivares-protegidas.

11. The adoption of GM seeds reached 91% of the soybean- and 81% of the corn-planted areas in 2013. CTNBio, which is the federal agency in charge of approving the development and commercialization of GMOs in Brazil, approved 38 genetically modified traits that can be used in plant seeds until 2013. Of those approved traits, only two were developed by EMBRAPA. The other 36 traits were developed by six multinational companies — Monsanto, Bayer, Syngenta, Dow, DuPont-Pioneer, and BASF.

12. This conversion of cerrado to agricultural use has not been without environmental costs, including loss of biodiversity, invasive species, soil erosion, and land degradation (see Klink and Machado, 2005).

13. Mueller and Martha Jr. (2008) provide an in-depth assessment of socioeconomic development of the cerrado since the 1970s.

14. EMBRAPA is a case of successful institutional innovation. Its success can be attributed to a number of factors: (i) sustained financial support from the federal government; (ii) organization as a public corporation that enables it to operate without the bureaucratic burden of public administration rules and independence from political influence; (iii) scale to achieve critical mass; (iv) decentralization in 45 research centers spread across the country; (v) development of human capital; and (vi) international engagement with research laboratories and technology transfer offices in five continents and agreements with higher education and research centers across the

world, including the USDA in the US, INRA and CIRAD in France, JICA and JIRCAS in Japan, and CGIAR (Alves, 2012). EMBRAPA is the central node of the National Agricultural Research System (SNPA) that includes 21 state-level research institutes and 144 higher-education institutions engaged in agricultural research projects across the country.

15. Brazilian government expenditure is organized into functions and programs. A function represents the higher level of aggregation of federal government expenses, including health, education, social security, and the two agriculture-related functions (agriculture and agrarian organization). A program consists of a group of government actions towards a specific policy goal (Gasques, 2006).

16. Agricultural support is defined as the annual monetary value of gross transfers to agriculture from consumers and taxpayers, arising from policies that support agriculture. Transfers included in the producer support estimate (PSE) are composed of market price support, budgetary payments, and the cost of revenue forgone by the government and other economic agents (OECD, 2014).

17. Gasques et al. (2012) found that the effect of investments in agricultural R&D on productivity growth in Brazilian agriculture is stronger than the impacts of rural credit availability and agricultural exports.

18. As many as 4 million farmers, or 78% of the total number of farmers in the country, did not receive any technical assistance in 2006 (IBGE, 2006).

19. According to the latest Census of Agriculture (IBGE, 2006), 2 million farmers (about 40% of the total number of farms) were not able to obtain rural credit in 2006.

REFERENCES

Alves, E., 2012. EMBRAPA: A successful case of institutional innovation. In: Martha Jr., G.B., Ferreira Filho, J.B. (Eds.), Brazilian Agriculture: Development and Changes. EMBRAPA, Brasilia, DF, pp. 143–160.

Chaddad, F.R., Jank, M.S., 2006. The evolution of agricultural policies and agribusiness development in Brazil. Choices 21 (2), 85–90.

Food and Agriculture Organization of the United Nations — FAO, 2008. Investing in Sustainable Agricultural Intensification: The Role of Conservation Agriculture. Rome.

Food and Agriculture Organization of the United Nations — FAO, 2010. Land Resource Potential and Constraints at Regional and Country Levels. Land and Water Development Division, Rome.

Food and Agriculture Organization of the United Nations — FAO, 2014. Aquastat Database. Available from: <www.fao.org/nr/aquastat>, (accessed 05.11.14.).

Faleiro, F.G., Farias Neto, A.L., 2008. Savanas: Desafios e Estratégias para o Equilíbrio entre Sociedade, Agronegócio e Recursos Naturais. Embrapa Cerrados, Planaltina, DF.

Fischer, G., Shah, M., 2010. Farmland Investments and Food Security, Report prepared under World Bank IIASA contract — Lessons for the large-scale acquisition of land from a global analysis of agricultural land use. International Institute for Applied Systems Analysis, Laxenburg, Austria.

Fischer, G., van Velthuizen, H., Shah, M., Nachtergaele, F., 2002. Global Agro-Ecological Assessment for Agriculture in the 21st Century: Methodology and Results. International Institute for Applied Systems Analysis, Laxenburg, Austria.

Gasques, J.G., Verde, C.M.V., Bastos, E.T., 2006. Gasto público em agricultura: retrospectiva e prioridades. Economia 7 (4), 209−237.

Gasques, J.G., Verde, C.M.V., Bastos, E.T., 2010. Gastos públicos em agricultura: uma retrospectiva. Revista de Política Agrícola 19, 74−90.

Gasques, J.G., Bastos, E.T., Valdes, C., Bachi, M., 2012. Produtividade da agricultura brasileira e os efeitos de algumas políticas. Revista de Política Agrícola 21, 83−92.

Instituto Brasileiro de Geografia e Estatística − IBGE, 2006. Censo Agropecuário, Brasilia, DF. Available from: <www.ibge.gov.br> (accessed 09.11.14.).

Klink, C.A., Machado, R.B., 2005. Conservation of the Brazilian cerrado. Conserv. Biol. 19 (3), 707−713.

Lal, R., 2008. Savannas and global climate change: source or sink of atmospheric CO_2. In: Faleiro, F.G., Farias Neto, A.L. (Eds.), Savanas: Desafios e Estratégias para o Equilíbrio entre Sociedade, Agronegócio e Recursos Naturais. Embrapa Cerrados, Planaltina, DF, pp. 81−102.

Mueller, C.C., Martha Jr., G.B., 2008. A Agropecuária e o desenvovimento socioeconômico recente do cerrado. In: Faleiro, F.G., Farias Neto, A.L. (Eds.), Savanas: Desafios e Estratégias para o Equilíbrio entre Sociedade, Agronegócio e Recursos Naturais. Embrapa Cerrados, Planaltina, DF, pp. 105−169.

Organization for Economic Co-operation and Development − OECD, 2005. OECD Review of Agricultural Policies: Brazil. ISBN 92-64-01254. 226 p.

Organization for Economic Co-Operation and Development − OECD, 2014. Producer and consumer support estimates, OECD agriculture statistics database. Available from: <http://dx.doi.org/10.1787/agr-pcse-data-en> (accessed 07.11.14.).

Sachs, J.D., 2001. Tropical Underdevelopment, Working paper 8119, National Bureau of Economic Research. Available from: <http://www.nber.org/papers/w8119>.

Spehar, C.R., 2008. Grain, fiber and fruit production in the cerrado development. In: Faleiro, F.G., Farias Neto, A.L. (Eds.), Savanas: Desafios e Estratégias para o Equilíbrio entre Sociedade, Agronegócio e Recursos Naturais. Embrapa Cerrados, Planaltina, DF, pp. 477−501.

CHAPTER 3

Agriculture in Southern Brazil: Cooperatives and Contract Farming

Contents

3.1 INTRODUCTION

Chapter 1 introduced the story of the Pinesso family and described how it evolved from peasant farmers working as sharecroppers in coffee plantations in Paraná to a large, commercial farming operation in the cerrado. Eugênio Pinesso, the son of Italian immigrants, Ferruccio and Anna Pinesso, who arrived in Brazil in 1923, worked on the farm since he was a kid. He did not have the opportunity to go to school but he was a hard-working, entrepreneurial, and shrewd businessman. He was an early adopter of new agricultural practices and technologies and was one of the first farmers to plant soybeans in Paraná in the late 1950s. With the use of lime and chemical fertilizers, Eugênio started to grow soybeans as the main crop in regions of the state with low natural soil fertility, where coffee plantations were not viable. By 1972 Eugênio was a well-established farmer with six farms in Paraná, totaling 1,500 hectares (ha). Despite his success in Paraná, Eugênio acquired 2,000 ha in the state of Mato Grosso do Sul (MS) in 1976. In 1983 Eugênio sold all his land in Paraná to

F. Chaddad: The Economics and Organization of Brazilian Agriculture.
DOI: http://dx.doi.org/10.1016/B978-0-12-801695-4.00003-3

acquire more land in Mato Grosso and the family moved to Campo Grande, MS. In 2014 the Pinesso Group planted 117,000 ha in the Brazilian cerrado and had more than 1,000 employees.

We will see later in this book how the Pinesso Group has expanded since the 1980s to become one of the major farming entities in the Brazilian cerrado. The story of the Pinesso family is not unique as the cerrado was conquered by the second and third generations of European immigrants who first established in southern Brazil in the first half of the twentieth century. But, unlike Eugênio Pinesso, many of these immigrants decided to stay in the south. In this chapter we will see how agriculture has developed in southern Brazil since the 1950s.

The Pinesso story is informative because it shows how the son of an Italian immigrant was able to make it as a farmer. Eugênio was not educated, but was extremely hard-working. He was ahead of his peers in adopting early several technologies developed by Brazilian researchers described in Chapter 2, such as soybean seeds adapted to local conditions, soil liming, and biological nitrogen fixation. He learned very early that cheap land with low natural fertility was not necessarily unproductive land. And he was entrepreneurial, uprooting his family from Paraná and moving to Mato Grosso in 1983.

Before continuing, I would like to pose the following hypothesis: *In addition to individual characteristics (and luck), the success of a farmer depends on access to basic factors of production (land, technology, and credit) and access to markets.* As noted in Chapter 2, Brazil has an abundance of land. But land availability alone is not enough to make it productive. Farmers must have secure property rights to land so that land can be bought, sold, and leased in well-functioning markets. Only when farmers have secure property rights to land will they bear the risks and make the necessary investments to make it productive. In addition, appropriate agricultural technology must be developed and disseminated for farmers to make land more productive, and credit must be available to enable them to adopt modern technologies such as improved seeds, fertilizers, crop protectants, and machinery. When these factors of production are available and used following appropriate agronomic practices — and, very importantly, Mother Nature cooperates — farmers are able to increase productivity and produce a surplus. But how does that surplus reach consumers in the cities and other countries? Farmers must have access to markets, which depends on infrastructure and vertically coordinated organizations — known as value chains or agrifood systems.

We learned in Chapter 2 that Brazil was successful in developing agricultural technologies adapted to the unique conditions of the cerrado. With the increased availability of subsidized rural credit dispensed by state-owned banks in the 1970s and 1980s, farmers were able to adopt these technologies to develop the cerrado, increase agricultural productivity, and transform Brazil from a food importer to the third-largest agricultural exporter in the world. Starting in this chapter we will see how different value chain configurations provided technology and credit to farmers — sometimes complementing but often substituting for the state in the absence of state-provided extension services and well-functioning rural credit markets — and linked Brazilian farmers to domestic and global markets. This chapter focuses on the southern region where cooperatives and contract farming[1] play a prominent role in linking farmers to markets.

3.2 THE DIVERSITY OF AGRICULTURAL ORGANIZATION

Agriculture in Brazil is very heterogeneous and diverse. Regional differences are immense (Table 3.1). According to the 2006 Census of Agriculture, there are 5.2 million farm establishments in Brazil. The average farm has 64 ha with gross farm income of BRL 35,350 (about US $17,000). The level of education is low — 24% of the principal farm operators are illiterate and only 10% have completed high school. Access to agricultural technology and rural credit is low — only 24% of the farmers reported receiving technical assistance and 18% obtained rural credit in 2006.

The northeast region has the greatest number of farms — 2.5 million — but they are smaller (31 ha), poorer (annual gross income of about BRL 13,000 per farm), and less educated (illiteracy rate of 41%) than the national averages (Table 3.1). Less than 9% of these farmers had access to technical assistance and only 13% were able to obtain rural credit in 2006. Access to technical assistance and credit are highly dependent on government programs such as PRONAF (Tables 3.2 and 3.3). Only 2% of farmers are members of a cooperative and less than 1% of them participate in a contract farming scheme. The northeast region has lagged behind in agricultural development. Farmers in the northeast were left behind and did not benefit from the modernization of Brazilian agriculture in the last four decades simply because they did — and continue to — not have adequate access to technology, credit, and markets. Farming continues to be a

Table 3.1 Characteristics of farm operators by region and in selected states

	Number of farms	Land in farms (ha)	Average farm size (ha)	Gross farm income (R$ million)	Average gross farm income (R$ per farm)	Share of farm operators who are illiterate (%)	Share of farm operators with high school diploma (%)	Share of farms that received technical assistance (%)	Share of farms that obtained rural credit (%)	Share of farms that are cooperative members (%)	Share of farms under contract farming (%)
Brazil	**5,175,636**	**333,680,037**	**64**	**163,986**	**35,350**	**24.5%**	**10.1%**	**24.0%**	**17.8%**	**10.6%**	**1.7%**
North	475,778	55,535,764	117	9,142	22,138	18.9%	6.4%	15.9%	8.9%	3.0%	0.9%
Northeast	2,454,060	76,074,411	31	29,219	13,128	41.0%	5.6%	8.6%	13.3%	1.8%	0.7%
Southeast	922,097	54,937,773	60	52,879	66,278	10.7%	18.0%	33.0%	15.2%	15.9%	1.3%
São Paulo	227,622	16,954,949	74	28,188	155,799	3.2%	30.3%	48.2%	13.3%	18.8%	2.1%
South	1,006,203	41,781,003	42	43,926	46,807	4.7%	12.9%	54.8%	36.5%	30.8%	4.9%
Paraná	371,063	15,391,782	41	16,735	49,468	6.2%	16.3%	49.6%	30.0%	26.9%	4.2%
Santa Catarina	193,668	6,062,506	31	9,035	50,635	2.7%	11.2%	61.4%	37.3%	25.9%	9.1%
Center-west	317,498	105,351,087	332	28,820	109,150	8.4%	19.5%	32.0%	14.0%	10.9%	1.5%
Mato Grosso	112,987	48,688,711	431	12,292	137,118	9.2%	14.6%	25.7%	13.9%	8.8%	1.3%

Source: IBGE (2006).

Table 3.2 Sources of technical assistance received by farmers by region and in selected states

	Government	Own/ hired	Cooperative	Integrator (contract farming)	Other
Brazil	**39.5%**	**20.1%**	**18.1%**	**12.4%**	**9.8%**
North	70.9%	17.8%	5.8%	1.5%	4.0%
Northeast	60.4%	25.1%	3.5%	2.5%	8.6%
Southeast	39.1%	28.6%	17.4%	4.4%	10.5%
São Paulo	31.5%	32.0%	16.8%	6.5%	13.2%
South	28.6%	11.1%	27.5%	23.4%	9.5%
Paraná	20.2%	13.4%	35.9%	17.2%	13.4%
Santa Catarina	35.6%	8.9%	17.5%	31.6%	6.3%
Center-West	33.7%	35.3%	9.0%	5.1%	16.9%
Mato Grosso	42.8%	30.1%	4.9%	6.6%	15.5%

Source: IBGE (2006).

Table 3.3 Sources of rural credit received by farmers by region and in selected states

	Share of farmers who needed credit but did not obtain (%)	Obtained rural credit from[a]:		
		Banks[b]	Credit cooperatives	Agribusiness[c]
Brazil	**48.1%**	**91.1%**	**7.0%**	**3.4%**
North	53.9%	93.7%	2.8%	1.2%
Northeast	59.0%	95.5%	2.0%	0.4%
Southeast	34.1%	91.0%	7.5%	1.2%
São Paulo	21.1%	91.4%	7.3%	1.6%
South	27.8%	86.8%	12.0%	7.0%
Paraná	31.0%	84.7%	13.7%	5.1%
Santa Catarina	23.9%	88.8%	9.6%	7.1%
Center-west	41.7%	92.1%	4.0%	4.6%
Mato Grosso	50.0%	87.5%	6.2%	7.1%

[a]The numbers in each row do not add up to 100% because a farmer may obtain credit from multiple sources in the same year.
[b]Unfortunately, the Census of Agriculture does not collect data about the type of bank conceding credit to farmers — in particular, state-owned vs. commercial banks. But the bulk of the subsidized rural credit is provided by state-owned banks such as Banco do Brasil.
[c]Agribusiness includes farm input suppliers, farm machinery, and equipment dealers and processors (integrators).
Source: IBGE (2006).

subsistence activity with low use of modern inputs and technology. Not surprisingly, the region only produces 18% of the total value of farm production and 10% of the national production of grains and oilseeds despite having almost 50% of all farm establishments in the country.

The southern region, in turn, is largely a success story. With about 1 million establishments farming 12.5% of the country's agricultural land, it produces 40% of the domestic grain and oilseed crop and 27% of the gross value of agricultural production. Farms are small, with an average area of 42 ha, but very successful in creating economic value from the land, with an average gross production value of about BRL 47,000 (US $21,000) per farm in 2006 — more than three times their peers in the northeast. They are more educated than the national average, but what really stands out in Table 3.1 is that 55% of the farm establishments in the southern region received technical assistance and 37% obtained rural credit in 2006, roughly double the rate of the respective national averages.

A distinguishing characteristic of agriculture in southern Brazil is the dominant role of cooperatives and contract farming arrangements in providing access to technology, credit, and markets for small, family farms. About 31% of farmers are members of a cooperative and 5% have contracts with downstream processors — primarily poultry and pork. As a result, they are less dependent on government programs to be successful. Table 3.2 shows that more than 50% of the farmers in the south received technical assistance from cooperatives and processors, compared to 28% from state organizations. Of those farmers who were able to obtain credit in 2006, 19% got it from cooperatives and processors. In what follows, we take a closer look at the organization of agricultural value chains in southern Brazil, focusing on the role of cooperatives and contract farming arrangements.

3.3 AGRICULTURE IN PARANÁ

As illustrated in the story of the Pinesso family, the development of agriculture in southern Brazil cannot be explained without the role of millions of European immigrants — Italians, Portuguese, Spaniards, Germans, Dutch, Ukrainians, Polish, etc. — that arrived in Brazil between the 1870s and the 1950s.[2] Most of them found work as sharecroppers on coffee plantations, just like the Pinesso family. Some were settled in colonies — groups of immigrants of the same ethnicity organized as a cooperative society in their home country that came to southern Brazil to develop a settlement of their own around agriculture. Such was the case of 50 Dutch families organized as the Emigration Cooperative Society in Hoogeveen, the Netherlands, that arrived in Castro, PR, in December 1951 and the group of 500 Swabian families from the former Austro-Hungarian Empire that arrived in Guarapuava, PR, to form the Entre Rios Colony, also in 1951.

These European immigrants and their offspring helped transform Paraná from its economic dependence on coffee plantations to a diversified agricultural powerhouse. Paraná is today a major agricultural state in Brazil — the number-one producer of barley, beans, corn, poultry, and wheat; the second largest producer of cassava, oats, rye, and soybeans; and a major producer of coffee, milk, pork, potatoes, tobacco, sugarcane, onions, and peanuts (Table 3.4). All of this production comes from 370,000 producers in 15 million hectares, which is equivalent to less than 5% of the country's land in farms (Table 3.1). The average size of a Paraná farm is 41 ha, with about BRL 49,500 (US$22,500) in annual gross farm income.

The successful development of agriculture in Paraná can be largely attributed to the strength of its farmer-owned cooperative organizations. According to 2013 data from the Paraná Cooperative Organization (OCEPAR), the state has 77 agricultural cooperatives with BRL 38 billion (US$17 billion) in total revenues, a combined market share of 56% in marketing the state's agricultural production (Table 3.5), and US$3 billion in agricultural exports. These cooperatives count 136,000 farmers as member-owners or one-third of all farms in the state. About 70% of these cooperative members farm less than 50 ha. Fifteen of the top-30 agricultural cooperatives in Brazil are based in Paraná (Table 3.6). These cooperatives are multipurpose because they provide a bundle of services to their farmer-members, including technical assistance, rural credit, farm input supply, storage and handling, processing, marketing, risk management, and exporting.

3.3.1 OCB and OCEPAR

Before looking at a few examples to illustrate the role of agricultural cooperatives in the organization and success of agriculture in Paraná, we must first understand the institutional context of cooperatives in Brazil. Up until 1966, cooperative leaders had considerable flexibility to form and organize cooperative associations in Brazil. The first law mentioning

Table 3.4 Agricultural and livestock production and productivity in Paraná (2012)

	Brazil			Paraná (PR)			PR share of national production (%)	PR ranking[a]
	Area (ha)	Production (ton)	Productivity (kg/ha)	Area (ha)	Production (ton)	Productivity (kg/ha)		
Barley	102,289	258,531	2,527	52,402	167,883	3,204	64.9%	1
Beans	2,726,932	2,821,405	1,035	471,109	701,952	1,490	24.9%	1
Corn	14,225,998	71,296,478	5,012	2,998,246	16,571,751	5,527	23.2%	1
Poultry	—	13,058,000	—	—	3,703,249	—	28.4%	1
Wheat	1,891,475	4,380,256	2,316	781,861	2,107,515	2,696	48.1%	1
Cassava	1,703,733	23,414,267	13,743	172,834	4,000,048	23,144	17.1%	2
Oats	180,245	392,367	2,177	68,259	171,813	2,517	43.8%	2
Rye	2,574	4,190	1,628	1,116	2,066	1,851	49.3%	2
Soybeans	24,937,814	65,700,605	2,635	4,454,418	10,925,878	2,453	16.6%	2
Coffee	2,092,582	3,061,513	1,463	67,070	91,897	1,370	3.0%	3
Milk	—	32,304,000	—	—	3,968,510	—	12.3%	3
Pork	—	3,238,000	—	—	529,700	—	16.4%	3
Potatoes	130,404	3,496,166	26,810	28,690	769,322	26,815	22.0%	3
Tobacco	405,488	806,685	1,989	69,402	156,495	2,255	19.4%	3
Sugarcane	9,407,078	670,757,958	71,304	649,965	43,882,516	67,515	6.5%	4
Onion	58,496	1,444,146	24,688	6,653	136,723	20,551	9.5%	5
Peanuts	106,223	328,154	3,089	2,606	6,129	2,352	1.9%	5

[a]Relative to other states in Brazil.
Sources: IBGE, ABIPECS (2014), and UBABEF (2014).

Table 3.5 Market share of cooperatives relative to total production in Paraná (2012)

	Cooperative share (%)
Rice	10
Oats	37
Coffee	35
Sugarcane	12
Barley	77
Bean	11
Cassava	5
Corn	61
Soybeans	74
Wheat	64
Milk	40
Poultry	28
Pork	34

Source: OCEPAR (2014).

cooperatives in Brazil dates back to 1903, but it was only with decree 22.239 of 1932 that the federal government set rules concerning the organizational characteristics of a cooperative. These rules and doctrines closely followed the Rochdale principles of cooperation, including open and voluntary membership, democratic control, service at cost, and limited return on capital.[3] In addition, cooperatives were granted favorable tax treatment relative to for-profit business enterprises. This initial phase of cooperative development in Brazil was liberal regarding the formation and functioning of cooperatives.

The subsequent development of agricultural cooperatives in Brazil was significantly affected by increased federal government interference − in both monitoring cooperatives and regulating agricultural markets − from the mid-1960s to the late 1980s. As discussed in Chapter 2, this period was characterized by "massive" federal government intervention in agriculture with subsidized rural credit, commodity price support, and regulated markets. Decree-law 59 of 1966 instituted a phase of increased federal intervention in cooperatives that lasted until 1988. This phase also saw the enactment of national law 5764 in 1971, which established the institutional framework within which the Brazilian cooperative system still operates today. The 1971 law defined the legal status of cooperatives and set well-defined rules for their organization and functioning. With articles 92 through 94 of the 1971 cooperative law, the Brazilian

Table 3.6 Largest agricultural cooperatives in Brazil (2013)

	Cooperative	State	Foundation year	Revenues (R$ million)	Ownership	Number of members
1	Copersucar S.A.	SP	1959	23,153	Corporation[a]	43
2	Coamo	PR	1970	7,779	Local cooperative[b]	26,276
3	Aurora	SC	1969	5,131	Central cooperative[c]	12
4	C. Vale	PR	1963	4,113	Local cooperative	14,600
5	Lar	PR	1964	2,644	Local cooperative	9,199
6	Cocamar	PR	1963	2,531	Local cooperative	11,800
7	Comigo	GO	1974	2,383	Local cooperative	6,220
8	Agrária	PR	1951	2,197	Local cooperative	600
9	Itambé S.A.	MG	1949	2,054	Corporation[d]	31
10	Cooxupé	MG	1957	2,016	Local cooperative	12,000
11	Copacol	PR	1963	1,964	Local cooperative	5,000
12	Cooperalfa	SC	1967	1,729	Local cooperative	15,000
13	Integrada	PR	1995	1,712	Local cooperative	7,429
14	Castrolanda	PR	1951	1,669	Local cooperative	782
15	Coopavel	PR	1970	1,554	Local cooperative	3,400
16	Frimesa	PR	1977	1,496	Central cooperative	5
17	Coopercitrus	SP	1964	1,462	Local cooperative	18,000
18	Batavo	PR	1941	1,398	Local cooperative	742
19	Copagril	PR	1970	992	Local cooperative	4,514
20	Cocari	PR	1962	977	Local cooperative	6,404
21	Cotrijal	RS	1957	972	Local cooperative	5,490
22	Coasul	PR	1969	935	Local cooperative	5,374
23	Copercana	SP	1963	912	Local cooperative	6,000
24	Cooperfibra	MT	2001	875	Local cooperative	197
25	Cotrisal	RS	1957	828	Local cooperative	9,490
26	Coacen	MT	2005	760	Local cooperative	137
27	Capal	PR	1960	758	Local cooperative	1,566
28	Cosuel	RS	1947	736	Local cooperative	3,600
29	Coplacana	SP	1948	734	Local cooperative	9,000
30	Copercampos	SC	1970	717	Local cooperative	1,189

[a]Copersucar S.A. is 100% owned by the Copersucar cooperative, which is owned by 43 sugarcane processors.
[b]A local cooperative is owned directly by farmer-members.

cooperative system lost its independence as the federal government reserved the right to oversee the organization and functioning of all types of cooperatives in the country. Between 1966 and 1988, a state agency known as INCRA (Instituto Nacional de Colonização e Reforma Agrária) regulated and controlled agricultural cooperatives.

The 1971 law also provided official legitimacy to the unification of the cooperative system based on sole representation by the Brazilian Cooperative Organization (OCB). Although it introduced major government interference, the law was a watershed for the cooperative movement. It provided the legal basis for the OCB system to organize and become viable. Consequently, all cooperatives operating in Brazil began to follow and conform to a standard business model, enabling their economic expansion. Since 1971 OCB has been the highest agency for representing all cooperatives in the country. OCB is responsible for promoting and defending the cooperative system at all political and institutional instances. It also has the responsibility for preserving and enhancing the system, and for encouraging and advising the cooperative societies. Since its creation, OCB has assumed responsibility for organizing the cooperative movement, in order to strengthen and consolidate it. Today there are 27 state-level cooperative organizations[4] (such as OCEPAR in Paraná) representing 6,600 cooperatives in 13 sectors of economic activity, including 1,500 agricultural cooperatives (OCB, 2014).

The 1988 constitution introduced the principle of "self-management," as the federal government waived its constitutional rights to interfere in the formation, organization, and functioning of cooperatives. The National Cooperative Secretariat (Senacoop) — which replaced INCRA in controlling the cooperative movement — and the National Cooperative Council (CNC) discontinued their oversight role after the 1988 Constitution. As a result of this institutional change, and the economic liberalization and agricultural policy changes presented in Chapter 2, agricultural cooperatives started to face an increasingly liberal, unprotected market environment. Since the 1990s, OCB has played an increasingly important role in self-regulating the cooperative movement in Brazil. To fulfill this role, OCB implemented two programs in the 1990s. The first called for investment funds from the government to enable cooperatives to restructure and adapt to changing market conditions. The second program was designed to fund and provide cooperative education throughout the country. Following effective advocacy efforts by

OCB, the federal government instituted the Revitalization Program for Agricultural Cooperatives (Recoop) and the National Cooperative Education Service (Sescoop) in 1998. Sescoop is funded by a 2.5% levy on the payroll of all cooperatives in the country. Funds are disbursed by state-level organizations that aggregate demand for educational programs from their cooperative members. For instance, the budget of Sescoop in Paraná was BRL 31 million in 2014 and funded more than 5,500 capacity-building projects that reached more than 151,000 participants. These projects included workshops, training, leadership development, and short- and long-duration courses, and were attended by cooperative members, employees, managers, and board directors.

3.3.2 Coodetec

Back in the early 1970s, agricultural research in Brazil was conducted by only a few state-level research institutes such as IAPAR in Paraná and IAC in São Paulo. EMBRAPA was founded in 1974. There were just a couple of private seed companies operating in the domestic market and the availability of adapted seed varieties was minimal. Agricultural cooperatives in Paraná then decided to invest in R&D to complement the research efforts of these state institutes and to develop seeds adapted to the specific growing conditions of their territories.

In 1974 OCEPAR created a research department to house a genetic breeding program initially focused on wheat and soybeans. The breeding programs were funded by the Wheat Research Fund and the Soybean Research Fund based on contributions by Paraná cooperatives in proportion to their respective planted areas of the two crops. Two researchers were hired to provide leadership to the wheat and soybean breeding efforts. The first step was to build a germplasm with the collection of commercial seeds from other public research institutions and genetic banks in Brazil and abroad. This germplasm would provide the building block for the hybridization of different seeds to generate genetic variability to be tested in experimental fields. Plants with superior characteristics would then be selected for further experiments under different production conditions across the state in fields provided by the participating cooperatives. The seeds with superior performance — i.e., higher yields than the commercial varieties in use — would be called new varieties (or cultivars) to be registered with the Ministry of Agriculture prior to market introduction. These new varieties developed by OCEPAR researchers would then be multiplied and marketed by the agricultural cooperatives to their farmer-members in the state.

Developing new seed cultivars from scratch demanded time and patience. It took an average of 8 to 12 years for a new seed variety to be ready for commercialization and use, but this initial research effort required continued commitment and funding from the cooperatives. The payoffs started to appear in the early 1980s. One of the early commercial successes of the OCEPAR Research Department was the soybean variety called *Iguaçu*, which was introduced in 1984 and had a 10-year commercial lifecycle in the state with a planted area surpassing 6 million hectares. Because the cooperatives had funded the breeding program since its inception, they did not have to pay any royalties to multiply and market these new seed varieties to their members. In 1982, OCEPAR inaugurated its first research center in Cascavel, PR, and expanded the scope of its research efforts to include corn and cotton breeding programs also based on financial contributions from the participating cooperatives.

As was explained in Chapter 2, Brazil has introduced new laws and regulations since the 1990s that significantly affected the domestic seed market – including the Intellectual Property Law of 1996, the Plant Variety Protection Law of 1997, the Seed Law of 2003, and the Biosafety Law of 2005. This new institutional environment provided protection for private investments in agricultural R&D and allowed private firms to reap the rewards of innovation by collecting royalties and technological fees. This, in turn, attracted the entry of new players to the domestic seed market. Given these changes in the marketplace, OCEPAR decided to convert its research department into a cooperative structure called Coodetec in 1995.

Coodetec was initially organized as a tier-two, federated cooperative with 38 tier-one, local cooperatives in PR as member-owners. It started to operate as a commercial seed company, charging royalty fees from the use of its seed varieties and hiring a sales force to market its products. In 1997, the cooperative brought to market the first products with the Coodetec brand name – three soybean, one wheat, and one cotton variety. Revenues from royalty income and seed sales increased from BRL 2.5 million in 1995 to more than BRL 200 million in 2013. In 2003 Coodetec seeds had conquered a 26% national market share in wheat, 20% in soybeans, and 15% in cotton. By 2014, its membership base had expanded to other states, such as Rio Grande do Sul, Santa Catarina, and Goiás, reaching 185,000 farmers that were members of 32 local cooperatives. Its asset base included five research centers with more than 900 ha in experimental fields in four states and a staff of over 660 employees, including 14 researchers with graduate degrees. Between 1974 and 2013, Coodetec developed more than 200 new seed varieties for its cooperative members. Its soybean breeding program was particularly successful, reaching an average market share of 30% in the Paraná market since 1984.

But another structural break in the mid-2000s again changed the nature of competition in the domestic seed market – the approval of research, development, and marketing of genetically modified (GM) seeds. By the late 2000s, the domestic seed market of the main cash crops – corn, cotton, and soybeans – was dominated by GM seeds. Coodetec had started its agricultural biotechnology program in 1998 in collaboration with the Federal University of Viçosa (UFV). It also entered commercial licensing agreements with Monsanto, Dow, and BASF to use their biotechnology traits, such as resistance to herbicides and insects, in its seed varieties. However, it did not generate sufficient research dollars to compete with multinational corporations. Because of the increasing capital requirements and risks associated with biotechnology R&D to develop GM seeds, and given the fact that the domestic seed market was much more competitive in the 2010s than in the 1970s, the cooperatives decided to exit the seed industry and announced the sale of Coodetec – including its germplasm and plant-breeding assets built since 1974 – to Dow in July 2014.

3.3.3 Castrolanda

The first Dutch immigrants arrived in the Campos Gerais region, PR, and settled in the Carambeí colony in 1911. These pioneers formed the Batavo cooperative in 1925 to process and market milk and dairy products. Two other groups of Dutch immigrants also settled in the same region – the Castrolanda colony in 1951 and the Arapoti colony in 1960 – and also organized cooperatives – Castrolanda and Capal, respectively. These three agricultural cooperatives formed by Dutch immigrants are today among the largest in the country (Table 3.6), with combined revenues of BRL 3.8 billion and 3,000 members that farm about 390,000 ha. The productivity levels achieved by these farmers are above the national and state averages, primarily due to the services provided by their cooperatives. In what follows, we explain the evolution of one of these cooperatives and how it helped farmers settle, develop, and increase productivity in the Campos Gerais region.

In 1951, 44 farmers formed the Emigration Cooperative Society in Hoogeveen, the Netherlands, with two objectives: to organize the export of the emigrants' assets to Brazil and to acquire land in Brazil for their settlement. Between 1951 and 1954, about 50 families from different Dutch provinces arrived in Castro, PR, and formed the Castrolanda colony and cooperative. According to a recent account of the Castrolanda history (Kiers-Pot, 2001), 362 immigrants arrived in the colony with about 1,200 livestock (primarily milk cows), 12 tractors, one truck, several jeeps, equipment, and tools. They were small farmers in the Netherlands struggling to make a living after World War II. Pooling their assets, and with a 10-year credit line from the Paraná government, they initially acquired 5,612 ha. Each family received a plot of land ranging from 35 to 200 ha based on its equity contribution to the Emigration Cooperative Society. In the first years of the settlement, these immigrants organized and built a church, a school, a library, and a youth club. But it was not until the 1970s that the colony had access to electricity, telephone service, and paved roads.

Castrolanda is located on a plateau 160 km northwest of the state capital, Curitiba, in a region called Campos Gerais. Back in the 1950s, the region was not considered suitable for agriculture due to its shallow, sandy soils with low natural fertility that were prone to erosion. So initially these immigrants survived from the milk production of the cows they had imported from the Netherlands. In 1954 they organized a Livestock Producer Organization to develop a genealogical record of the dairy herd, to promote its genetic improvement via artificial insemination, and to market registered animals to other farmers. A Livestock Producer Training Center (CTP) was established in 1966 to develop human resources to work in the dairy farms and to transfer knowledge to producers outside the cooperative. Between 1966 and 2000, the CTP trained more than 10,000 farmers in best milk production practices. Producers also received technical assistance, credit, and farm inputs from the cooperative.

In 1957 Castrolanda formed a tier-two, federated cooperative with Batavo — the sister cooperative of Dutch immigrants located 30 km from Castrolanda — to build a milk-processing plant in Carambeí, PR. In its first year of operation, the plant processed 5 million liters of raw milk into pasteurized milk, cheese, butter, and other dairy products. By 1990, the region was considered one of the best milksheds in Brazil, with high productivity and milk quality levels, and the plant processed 85 million liters of raw milk. The Batavo-branded products reached consumers in the major Brazilian cities and competed favorably with other domestic and multinational processors. Following dramatic structural changes in the domestic milk market in the 1990s, the cooperatives decided to sell the processing plant and brand name in 1997. The cooperatives refocused their work on providing technical assistance and farm inputs to producers to help them increase productivity and quality. In addition, they formed a marketing pool — called Pool ABC[5] — to market the milk collectively to processors. This strategy worked well until the late 2000s, when the cooperatives decided to invest in a new processing plant. Milk volume processed in this new processing plant increased from 88 million liters in 2008 to 560 million liters in 2013. The plant sells dairy ingredients to food processors, and the cooperatives are investing in brand names linked to their Dutch heritage for consumer products distributed through retail.

Despite settling in a region with poor soils, and having accumulated knowledge and assets in milk production, Castrolanda also developed technologies to help their members diversify to other agricultural products. Between 1957 and 1979 the Dutch government funded one agronomist to develop crop production systems adapted to the Campos Gerais growing conditions and to provide technical assistance to farmers. This research and outreach effort paid off quickly and by 1968 Catrolanda farmers were no longer dependent on milk production. The crop production area of Castrolanda members increased from 2,000 ha in the early 1960s to 36,000 ha in 2000 and 125,000 ha in 2010. Current crop production includes soybeans, corn, wheat, potatoes, and other crops — in addition to milk and pork (Table 3.7). "Without the initial orientation of

Table 3.7 Agricultural and livestock production (tons) of Castrolanda members

	1965	1975	1985	1995	2000	2010
Soybeans		16,116	36,960	70,328	90,931	189,867
Corn			16,941	77,736	94,595	206,775
Wheat		5,215	5,052	9,694	21,629	107,686
Potatoes				1,801	2,100	45,931
Milk	5,506	7,638	19,202	54,646	81,086	166,299
Poultry[a]	24	1,838	8,300	22,800		
Pork	131	527	3,814	13,871	17,482	32,558

[a]The cooperative exited the poultry industry in 1997.
Source: Castrolanda.

these agronomists we would not have been able to develop agricultural production in the region. This was perhaps the greatest contribution of the Dutch to agriculture in Brazil," Frans Borg — the Board Chair of Castrolanda and son of one of the immigrant families — shared with me in a personal interview.

Mr. Hartman, the first agronomist sent by the Dutch government to work with Castrolanda farmers, set up an agricultural school in 1958 and the first experimental fields of the cooperative to test new seed varieties and production systems. Farmers formed a committee to oversee the experiments and identify the best crops adapted to the region. Mr. Hartman and his successors also fostered the development of producers' associations to share best practices and organized field days to allow private firms to showcase new technologies and farm machinery to farmers. According to the account of one Castrolanda pioneer, "we always tried to be as close as possible to the technical level of Dutch agriculture and we never saved efforts to provide the best education to our kids" (Kiers-Pot, 2001, p. 147). Since the 1970s, the cooperative made contributions to the Research Department of OCEPAR and started to produce its own seeds. With the accumulated knowledge developed by the Dutch agronomists, Castrolanda was able to organize its own Technical Department in 1979. It hired a team of agronomists formed in Brazilian schools to provide technical assistance to farmers. To this day Castrolanda farmers receive technical assistance services delivered by the cooperative Technical Department staff at no cost. Perhaps not surprisingly, the productivity levels achieved by Castrolanda farmers are significantly higher than the national and state levels. For example, the average productivity of the main crops in Castrolanda were the following in 2012: 3,659 kg/ha for soybeans, 10,852 kg/ha for corn, and 3,016 kg/ha for wheat, which are approximately 30% higher than the state and national averages for soybeans and wheat, and almost double for corn (Table 3.4).

The three cooperatives formed by Dutch immigrants — Arapoti (Capal), Batavo, and Castrolanda — innovated again in 1984 when they joined forces to create the ABC Foundation, focusing on agricultural R&D and outreach to farmers. The foundation is a private, non-profit organization funded by contributions from the producers of the three cooperatives and fees for services provided to other entities, such as soil testing and product testing in its experimental fields. In the 1980s, the ABC Foundation pioneered the development of no-till farming in Brazil ahead of the public research institutes. Given the low fertility and the high susceptibility to erosion of the Campos Gerais soils, no-till farming was a major technological innovation that required investments in new seeding equipment, weed control, and crop rotation. With the successful development of no-till farming, the ABC Foundation entered agreements with EMBRAPA, IAPAR, and other research and outreach institutes to disseminate the technology to other regions. As we saw in Chapter 2, Brazil is currently a world leader in the adoption of minimum tillage farming systems. To this day the ABC Foundation continues to engage in applied agricultural research in experimental fields located in four municipalities in Paraná and one in São Paulo. It has ongoing research partnerships with leading universities and research institutes in Brazil and has a team of researchers in the areas of soil fertility, crop science, weed science, plant pathology, entomology, agricultural mechanization, animal nutrition, pastures, agro-meteorology, and geographical information systems (GIS).

Given the increased agricultural production in the Campos Gerais region, spurred by the investments in technologies adapted to the local conditions since the 1950s, the three cooperatives increased their strategic alliance to co-invest in processing facilities. In addition to the dairy-processing plant described above, new investments include state-of-the art wheat milling, pork-processing, and feed-processing facilities. In addition to providing the technology and services necessary for farmers to adopt new technologies and increase productivity, the example of these cooperatives shows how they provide access to markets and opportunities for farmers to add value to their production. These cooperative strategies are essential to the economic viability and survival of small family farms in the region.

3.3.4 Agrária

In the 1750s, a group of successful south German farmers — the Swaben — relocated to the southeastern region of the Austro-Hungarian Empire to develop agriculture along the Danube River. They farmed there until 1944 when their assets were expropriated following the Russian invasion in WWII. They lived as refugees in Austria until 1951 when they received a grant from a humanitarian Swiss foundation to find land to settle in Brazil. A commission led by agronomist Michael Moor identified 22,000 ha of available land near the town of Guarapuava in the south-central region of Paraná to settle the group. About 500 Swaben families arrived in 1951, settled in the Entre Rios colony and formed the Agrária cooperative. Each family received 15 ha of land plus 8 ha for each son and 4 ha for each daughter.

In contrast to the Dutch immigrants, the Swaben arrived with no assets and had to start from ground zero to develop agriculture in order to repay the loan from the Swiss foundation. They also had no experience in livestock production. "Grains run in the veins of our members," according to Jorge Karl, the cooperative president, in a personal interview. So they first attempted planting wheat but with no success because the soils were not fertile. Only rice would grow in the region, albeit with low productivity. Between 1951 and 1966, the Swaben immigrants struggled and many families left the colony to find work in Curitiba or São Paulo. It was only in the late 1960s that agriculture started to develop in the region with the introduction of soybeans, investments in soil fertility, and adoption of appropriate agronomic practices.

Similarly to Castrolanda, the Agrária cooperative also invested in agricultural R&D and outreach to develop seeds and technologies for the specific growing conditions of its locale. As a member of OCEPAR, Agrária members contributed to the wheat and soybean research funds that led to the formation of Coodetec. The cooperative also organized its own research foundation, called FAPA, following the model of the ABC Foundation of the Dutch cooperatives. Agricultural productivity and production have soared since the 1970s and the cooperative grew horizontally until the mid-1990s. Current productivity levels for the main crops are as high as those achieved by Castrolanda farmers — 3,883 kg/ha for soybeans, 11,828 kg/ha for corn, and 2,733 kg/ha for wheat. Agrária producers reached 4,800 kg/ha in barley production in 2013, almost twice the national productivity average. The cooperative has been a regular recipient of awards due to the high productivity levels achieved by its farmer-members in the last 15 years.

The structural changes of the 1990s brought significant challenges to cooperatives in Brazil, and Agrária was not immune. Between 1995 and 1999 the cooperative incurred heavy losses and almost went bankrupt. But with new leadership, strategic repositioning, and strong member commitment, the cooperative was able to turn around. The major strategic change was to build vertically integrated value chains from experimental fields to processing for its main crops. Building on its agricultural expertise, Agrária invested in soybean-, corn-, wheat-, and barley-processing plants and became a major supplier of ingredients to major food and beverage companies. For example, Agrária supplies about 25%

of all the malting barley for beer production in Brazil. It developed a strategic alliance with German company IREKS to market specialty wheat flour products for bakeries in the domestic market. It also invested heavily in total quality and continuous improvement programs, including at the farm level. Finally, its strategic plan was tied to clearly defined performance measures in sustainability (zero accidents in its facilities), financial performance of the cooperative (profitability), and grower returns (annual increases of 5%). When it almost went bankrupt in 1999, the cooperative had sales of BRL 257 million and incurred net losses of BRL 108 million. In the last 3 years of operation (2012–2014), the cooperative had sales above BRL 2.1 billion and net income averaged BRL 74 million per year. As a result, the cooperative net worth was rebuilt from a negative value in 1999 to BRL 821 million in 2013.

The cooperative currently has 600 members that farm about 116,000 ha. These members are fully committed to Agrária, which means they have to acquire 100% of their credit needs and farm inputs from and deliver and market 100% of their crop production to the cooperative. In other words, all of their business transactions revolve around the cooperative. "Their livelihoods are in our hands," says Mr. Karl. This strong member commitment allows the cooperative to effectively link these producers to food ingredient buyers and maximize the value of their farm production. Because the value chains are coordinated and led by a cooperative, the value created returns to farmers as services (such as technical assistance and capacity building at no cost), higher prices for their commodities, and net worth accumulated in the cooperative (about BRL 1.3 million per member as of December 2014).

An alternative to the farmer-led, cooperative model is contract farming. For instance, AMBEV, the largest beverage company in Brazil, with sales of BRL 35 billion in 2013, acquires malting barley via production contracts with 2,200 family farms in southern Brazil — in addition to its supply agreement with Agrária. In the next section of this chapter, we analyze in more depth this alternative value chain configuration, which is also very common in southern Brazil.

3.3.5 Agricultural Cooperatives in Other Regions

Obviously there are agricultural cooperatives not only in the southern region, but across the country. According to national statistics collected by OCB in 2013, there are 1,516 agricultural cooperatives in Brazil, with more than 1 million members and 164,000 employees. The 30 largest agricultural cooperatives had combined revenues of BRL 77 billion in 2013 (Table 3.6). The market share of cooperatives relative to national production is impressive — 74% for wheat, 57% for soybean, 48% for coffee, 44% for cotton, 43% for corn, 35% for rice, and 30% for milk.

Most of these cooperatives operate similarly to the examples of Castrolanda and Agrária described above. They are multipurpose and provide several services to their members, including credit, technical assistance, farm inputs, storage and handling, marketing, and processing. Like Castrolanda and Agrária, some cooperatives are organized by groups of the same ethnicity. But the majority are organized by farmers in the same region such as Coamo, CVale, Lar, and Cocamar in Paraná, Comigo in Goiás, Cooperalfa in Santa Catarina, and Cooperfibra and Coacen in Mato Grosso (Table 3.6). Some cooperatives also tend to focus on one commodity, which is the case in the coffee and milk sectors.[6]

Brazil also has a few successful tier-two, federated cooperatives — known as "central cooperatives" — as these are cooperatives whose members are other cooperatives. These central cooperatives tend to focus on the processing and marketing of a few commodities with brand names. Examples of such central cooperatives include Aurora (pork, poultry, and milk processing), Itambé (milk processing), and Frimesa (pork and milk processing). More recently, central cooperatives have been formed to gain economies of scale and bargaining power in the import and marketing of fertilizers and crop protectants. Examples include Coonagro in Paraná and Coabra in Mato Grosso. Chapter 5 discusses the role of new-generation cooperatives in the agricultural frontier of Mato Grosso.

3.4 AGRICULTURE IN SANTA CATARINA

Santa Catarina (SC) is a small state tucked between Paraná and Rio Grande do Sul in southern Brazil. The state has about 193,000 farm establishments on 6 million hectares, which is less than 2% of the country's farmland (Table 3.1). Santa Catarina farms are half the size of the national average (31 ha) but producers are able to generate 40% more gross income per farm than the national average (more than BRL 50,000). Farmers in SC have the highest levels of access to technical assistance (61% of the farms) and rural credit (37% of the farms), received primarily from cooperatives and agribusinesses in contract farming arrangements. About 26% of the farmers in the state belong to a cooperative and 9% participate in long-term contractual arrangements with processors.

Despite its relatively small size, the state has a diversified agricultural production and is a major producer of apples, onions, pork, garlic, poultry, rice, tobacco, and other crops and livestock products (Table 3.8). It is the largest pork producer in the country and the second largest poultry producer, behind Paraná. Similarly to Paraná, cooperatives play a significant economic role in the state. There are 52 agricultural cooperatives with 67,500 members, 32,000 employees, combined revenues of BRL 13 billion, and combined net worth of BRL 3.3 billion, according to 2014 data from the state cooperative organization (OCESC, 2014). The third largest agricultural cooperative in Brazil is headquartered in Chapecó, SC, located 550 km to the west of the state capital, Florianópolis. Aurora is a central cooperative owned by 12 local cooperatives in SC and three other states, which together have a membership of about 60,000 farmers (Table 3.6).

Table 3.8 Agricultural and livestock production and productivity in Santa Catarina (2012)

	Area (ha)	Production (ton)	Productivity (kg/ha)	SC share of national production (%)	SC ranking[a]
Apples	18,493	530,601	28,692	49.3%	1
Onions	18,942	376,603	19,882	28.0%	1
Pork	–	805,500	–	24.9%	1
Garlic	1,908	19,315	10,123	17.8%	2
Poultry	–	2,347,828	–	17.9%	2
Rice	149,127	1,020,111	6,841	9.6%	2
Tobacco	117,060	244,458	2,088	29.6%	2
Bananas	29,935	683,592	22,836	10.1%	3
Wheat	66,591	139,428	2,094	3.2%	3
Grapes	–	70,909	–	4.9%	4
Milk	–	2,717,860	–	8.4%	5
Potatoes	5,324	115,924	21,774	3.6%	5
Tomatoes	2,496	169,826	68,039	4.6%	7
Beans	80,348	135,868	1,691	4.1%	7
Corn	484,420	3,326,284	6,867	4.0%	8
Soybean	521,339	1,586,351	3,043	1.6%	12
Cassava	26,838	506,906	18,888	2.2%	13

[a]Relative to other states in Brazil.
Sources: IBGE, ABIPECS (2014), and UBABEF (2014).

Aurora processes about 10% of the pork and 4% of the poultry produced in the country and markets more than 800 branded food products in Brazil and abroad. Aurora and the other meat-processing cooperatives in southern Brazil compete with the three investor-owned giants in the sector — Brasil Foods, JBS, and Marfrig — which are among the largest meat processors in the world.

Western Santa Catarina is the birth place of contract farming arrangements for poultry and pork production in Brazil. Contract farming arrangements were introduced by meat processors back in the 1960s as a means to transfer technology to growers, foster production growth, and thereby assure supply. Contracts allow closer coordination between farmers and buyers (processors and retailers) than spot markets and in general provide the buyer more control over production decisions at the farm level. For this reason, buyers of agricultural produce engaged in contractual agreements with farmers are known as "integrators." The adoption of these arrangements has grown over time and by the 1990s they were the dominant form of organization in the pork and poultry supply chains at the expense of independent producers selling their finished livestock in spot markets. Today, more than 95% of pork and poultry production in Brazil is carried out under contract farming schemes. Both cooperatives and investor-owned firms rely on contracts with hog and chicken growers, although there are differences in how these contracts are set up, which affects the relationship between suppliers and integrators and performance.[7]

The characteristics of these contracts have changed over time but today there are two basic types of contract — marketing contracts and production contracts.[8] In a marketing contract, the producer retains all production-related decisions — such as choice of major inputs (genetics, feed, and veterinary products) — and commits part or the entirety of his production to a processor to be delivered at a future date at a fixed price. The marketing contract specifies the price (or a pricing formula), the quantity to be delivered, and the delivery outlet. The farmer retains ownership of the livestock until delivery to the buyer and has substantial autonomy over production decisions, while shifting some of the commodity price risk to the buyer.

In the case of the production contract, the integrator provides the farmer technical assistance, credit, and the input package (genetics, feed, and veterinary products). The farmer, in turn, provides labor, electricity, and production facilities such as housing for the livestock and the necessary equipment to grow them. The finished hogs or broilers are picked up and transported to the processing plant by the integrator, assuring the grower of a market. The major difference is that under the production contract, the integrator retains ownership of the livestock and the producer plays the role of a service provider. In addition, the integrator has autonomy to set quality standards to producers, including intrinsic meat quality attributes and conformity with sanitary and animal welfare requirements. For example, the use of growth hormones is not allowed in broiler production and growers must adopt industry best practices regarding animal welfare and waste disposal. If the grower does not achieve minimum quality levels set by the integrator, the price will be discounted relative to the reference contract price or the contract might be terminated. In other words, the grower shifts price risk to the processor, has access to credit, technology, and markets, but at the expense of autonomy and control over major production decisions.

Under a production contract, the producer compensation is determined by a pricing formula set by the integrator based on production efficiency criteria and conformity with best production practices. Most integrators set their pricing formulas based on the relative production efficiency achieved by growers – a "meritocracy" system in which producers with higher performance levels are paid more. For example, the larger the average weight of finished livestock, the higher the survival rate of the batch, and the lower the feed conversion ratio (i.e., the amount of feed converted into meat), the greater the compensation of the grower. Production contracts also specify the discount schemes for low quality. A grower will receive lower prices, sometimes not enough to cover production costs, if he performs below average. This payment system, based on relative performance, effectively weeds out producers who cannot achieve above-average performance on a continuous basis or are unable to follow the increasing production conformity and product quality levels set by the processors.

The increased use of production contracts has been associated with production growth, increasing productivity, and higher meat quality levels. As we saw in Chapter 1, meat production in Brazil has soared from 3.4 to 24.6 million tons between 1975 and 2010. Poultry production has grown 1,400% since 1960, when it was virtually non-existent, and surpassed 13 million tons in 2011. Brazilian poultry exports started in 1976 and increased from 300,000 tons in 1990 to 3.9 million tons in 2011, reaching more than 150 countries. Brazil has become the world's leading exporter and third largest producer of poultry.

Similarly to poultry, Brazilian pork production increased from 2.8 million tons in 2002 to 3.5 million tons in 2012, primarily based on productivity gains. During this period, the number of finished hogs per sow increased from 18 to 23 and the average carcass weight increased from 77 to 88 kg. As a result of productivity gains and meat quality improvements, Brazil was able to conquer international markets with annual exports exceeding 500,000 tons since 2010, compared to 128,000 tons in 2000. The country is now the fourth largest exporter of pork with about 8% of the world's market share. The international competitiveness achieved by the Brazilian pork and poultry producers is largely the result of research conducted by the EMBRAPA Hogs and Poultry center located in western Santa Catarina — in animal genetics, feed composition, disease prevention and control, and waste disposal — and the tight coordination of the supply chain via production contracts.

But at the same time that the production and exports of pork and poultry increased, the sectors went through significant structural changes, especially since the 1990s. Processors have consolidated and the industry structures in both poultry and pork are now oligopolies. The four largest firms process 88% of all hogs in the country, while the two largest firms export 70% of all poultry. The adoption of production contracts between integrators and growers with performance-based pricing formulas has led to increased grower concentration. For example, pork production in SC has increased from 230,000 tons in 1985 to 682,000 tons in 2006 but the number of producers has decreased from 35,000 to 12,000 in the same period. Not surprisingly, farmers have several grievances with integrators regarding the quality of the production inputs received, lack of transparency about pricing data, and the continuous requirements for new investments to update facilities, increase production, and conform to more stringent quality levels.

In sum, pork and poultry production is organized by large integrator companies and cooperatives that provide genetics, feed, veterinary services, and technical assistance to small-scale, family farmers. The integrators have production contracts with growers to raise the hogs and the broilers. Integrators vertically integrate into livestock breeding and feed production, which provides them control of the relevant production inputs. Vertical integration — in association with animal confinement — enables integrators to effectively manage the production risk associated with broiler and hog production and thereby assure a dependent supply of finished animals for slaughtering and processing. In addition, vertical coordination of the supply chain with production contracts allows integrators to monitor production practices and meat quality, and to provide information and technical assistance to growers. Monitoring and performance-based compensation are mechanisms used to measure production efficiency, incentivize growers to increase productivity, and to assure grower compliance with quality requirements. With effective organization of the supply chain, integrators are able to guarantee the traceability of the final product from the farm to the table of consumers.

3.5 SUMMARY

In this chapter we advanced the hypothesis that *in addition to individual characteristics (and luck), success of a farmer depends on access to basic factors of production (land, technology, and credit) and access to markets.* We saw that farmers in the northeastern region were left behind and did not benefit from the tremendous productivity gains and development of Brazilian agriculture since the 1970s precisely because they did not have adequate access to technology, credit, and access to markets in an organized way.[9] Their counterparts in southern Brazil fared much better. We explained that this is so because farmers were organized in well-functioning value chains that provide them access to the production inputs, technology, credit, and information necessary to increase productivity and thus production. In addition, these well-organized and tightly coordinated value chains effectively link agricultural production from the farms to the tables of consumers in Brazil and abroad.

This chapter focused on two alternative value-chain configurations linking family farmers in the southern region to markets — agricultural cooperatives and contract farming arrangements with meat processors. The examples presented in this chapter provide clear evidence that both arrangements facilitated farmers' access to technology, credit, and markets. The resulting productivity gains allowed these value chains to gain international competitiveness, in addition to increasing food availability and decreasing prices in the domestic market. However, the roles played by farmers in these arrangements differ markedly, as well as their decision-making autonomy and ability to capture a share of the total economic value created. In the following chapter we analyze the organization of value chains in the southeastern region, with a focus on two export-oriented sectors — sugarcane and orange juice.

NOTES

1. Contract farming is an arrangement where an agribusiness — often a food processing company or retailer — engages on a long-term basis with family farms that grow and supply the agribusiness with the raw material that it needs (such as chicken and hogs for meat processing, milk for dairy processors, tomatoes for a canning factory, malting barley for breweries, and fresh fruits and vegetables for a supermarket chain). In general, these arrangements are based on marketing agreements that establish pricing formulas, quality standards, and timing for the raw material delivery. Often, these marketing agreements are complemented with production contracts where the agribusiness provides farm inputs, credit, and technical assistance to the farmer, who in turn provides the necessary facilities and labor to grow the product delivered to the agribusiness. For additional information and examples, refer to FAO (2013).
2. Most of these immigrants, about 75,000 per year, arrived between 1875 and 1930. Immigration to Brazil was principally from Europe, but the country also received immigrants from Japan, Syria, Lebanon, and other countries.
3. The Rochdale principles are a set of ideals for the operation of cooperatives. They were developed by the Rochdale Society of Equitable Pioneers in Rochdale, England, in 1844 and were officially adopted by the International Cooperative Alliance in 1937. The Rochdale principles provide the basis for the operation of cooperatives around the world to this day.
4. These state-level organizations (known as OCEs) are members of the OCB system with the responsibility for registering, guiding, and integrating cooperatives; promoting training and capacity building; and enabling professionalism and self-management of cooperatives in each state.
5. The marketing pool ABC combined milk producers from the three cooperatives — Arapoti (Capal), Batavo, and Castrolanda.
6. For the interested reader, I have described the role of cooperatives in the coffee and milk sectors in other publications (Chaddad and Boland, 2009; Chaddad, 2014).
7. In a recent paper, Cechin et al. (2013) analyzed broiler contracts in Paraná and found significant differences in the relationship characteristics between growers and the integrator. These relationship differences, in turn, explain why suppliers delivering to a cooperative have a better performance regarding product quality than suppliers of corporate integrators.
8. These contractual arrangements are very similar in Brazil and the US. For details, refer to MacDonald and Korb (2011) and Miele and Miranda (2013).
9. Other social scientists might argue that underdevelopment in the Brazilian northeast is also explained by historical, institutional, and geographic factors. I do not disagree with them.

REFERENCES

Associação Brasileira da Indústria Produtora e Exportadora de Carne Suína — ABIPECS, 2014. Estatísticas. Available from: <www.abipecs.org.br/pt/estatisticas.html> (downloaded 05.12.14.).

Cechin, A., Bijman, J., Pascucci, S., Zylbersztajn, D., Omta, O., 2013. Quality in cooperatives versus investor-owned firms: evidence from broiler production in Paraná, Brazil. Manage. Decis. Econ. 34, 230—243.

Chaddad, F.R., 2014. Responding to the external environment: the evolution of Brazilian dairy cooperatives. In: Mazzarol, T., Reboud, S., Limnios, E., Clark, D. (Eds.), Research Handbook on Sustainable Cooperative Enterprise: Case Studies of Organizational Resilience in the Cooperative Business Model. Edward Elgar Publishing, Cheltenham, UK, pp. 100–112.

Chaddad, F.R., Boland, M., 2009. Strategy-structure alignment in the world coffee industry: the case of cooxupé. Rev. Agric. Econ. 31 (3), 653–665.

Food and Agriculture Organization – FAO, 2013. Contract Farming for Inclusive Market Access, Rome. Available from: <www.fao.org/publications> (accessed 15.01.15.).

Instituto Brasileiro de Geografia e Estatística – IBGE, 2006. Censo Agropecuário, Brasilia, DF. Available from: <www.ibge.gov.br> (downloaded 09.11.14.).

Kiers-Pot, C.H.L., 2001. Castrolanda 50 Anos: 1951–2001. Kugler Ltda, Castrolanda, PR.

MacDonald, J., Korb, P., 2011. Agricultural contracting update: contracts in 2008. Economic Information Bulletin No. 72. U.S. Department of Agriculture, Washington, DC. Available from: <www.ers.usda.gov/publications/eib-economic-information-bulletin/eib72.aspx>.

Miele, M., Miranda, C.R., 2013. O Desenvolvimento da Agroindústria Brasileira de Carnes e as Opções Estratégicas dos Pequenos Produtores de Suínos do Oeste Catarinense no Início do Século 21. In: Campos, S.K, Navarro, Z. (Eds.), A Pequena Produção Rural e as Tendências do Desenvolvimento Agrário Brasileiro: Ganhar Tempo é Possível? CGEE, Brasília, DF, pp. 201–232.

Organização das Cooperativas Brasileiras – OCB, 2014. Dados do Cooperativismo Brasileiro. Available from: <www.ocb.coop.br> (accessed 10.12.14.).

Organização das Cooperativas do Paraná – OCEPAR, 2014. Dados do Cooperativismo Paranaense. Available from: <www.paranacooperativo.coop.br> (accessed 10.12.14.).

Organização das Cooperativas de Santa Catarina – OCESC, 2014. Dados do Cooperativismo Catarinense. Available from: <www.ocesc.org.br> (accessed 10.12.14.).

União Brasileira de Avicultura – UBABEF, 2014. Annual Report 2014. Available from: <http://www.ubabef.com.br/publicacoes> (accessed 05.01.15.).

CHAPTER 4

Agriculture in Southeastern Brazil: Vertically Integrated Agribusiness

Contents

4.1 INTRODUCTION

In the previous chapter we showed the diversity of agricultural organization in Brazil and discussed the role of cooperatives and contract farming arrangements in the southern region. In this chapter we continue our analysis of alternative forms of agricultural organization, focusing on the role of vertically integrated agribusiness in southeastern Brazil. A vertically integrated agribusiness has operations in several stages of the value chain — for example, a sugar producer that grows the sugarcane and then processes it into sugar. The southeastern region is the most developed in the country and performs better than the national averages with respect to income per capita, education, infrastructure, and other human development measures. The region is highly urbanized, with the largest

F. Chaddad: The Economics and Organization of Brazilian Agriculture.
DOI: http://dx.doi.org/10.1016/B978-0-12-801695-4.00004-5

metropolitan areas in the country, including São Paulo, Rio de Janeiro, and Belo Horizonte.

The southeastern region is comprised of four states: São Paulo, Rio de Janeiro, Minas Gerais, and Espírito Santo. The states of São Paulo and Minas Gerais have sizable agricultural sectors with combined gross farm income of BRL 49 billion or 30% of the nation's total in 2006 (Table 4.1). Minas Gerais is the state with the largest number of farms in the country. It is home to more than 550,000 farm establishments, which is about 10% of all farms in Brazil. The state has a diversified agricultural sector and is the largest producer in the country of coffee and milk, and a major producer of beef, sugarcane, potatoes, beans, corn, fruits, and vegetables. A perusal of Table 4.1 shows that farms in Minas Gerais are not significantly different from the national averages. The organization of agriculture in the state is very diverse and no clear pattern is observed. As discussed in the previous chapter, cooperatives play a significant role in the coffee and milk sectors but they are not the dominant form of organization in these value chains. The largest coffee (Cooxupé) and the largest dairy cooperative (Itambé) in the country are headquartered in Minas Gerais.

Agriculture in the state of São Paulo, on the other hand, has some unique characteristics. The 230,000 producers in the state farm about 17 million hectares, which is equivalent to 5% of the total farmland in the country (Table 4.1). But the state gross value of production surpasses BRL 28 billion or 17% of the country's total. Farms in São Paulo are slightly larger in size (74 ha) compared to the national average of 64 ha, but the average gross farm income reaches BRL 156,000 per farm, almost fivefold the national average of BRL 35,000. Farmers in the state have significantly higher levels of education and better access to technical assistance than their peers across the country.

From the 1880s and up until the 1930s, coffee was the most important crop in the country and São Paulo was the hub of the coffee industry as the largest producer. It attracted many European immigrants such as the Pinesso family described in Chapter 1. Coffee export earnings provided the economic impetus for the first wave of industrialization in the state. But after the 1929 crash and economic crisis, many coffee plantations were eradicated and replaced by pastures and other crops. Today coffee plantations cover only 1% of the agricultural land in the state, primarily in municipalities near the Minas Gerais border. Since the 1970s sugarcane has become the primary crop in São Paulo and

Table 4.1 Characteristics of farm operators by region and in selected states

	Number of farms	Land in farms (ha)	Average farm size (ha)	Gross farm income (R$ million)	Average gross farm income (R$ per farm)	Share of farm operators who are illiterate (%)	Share of farm operators with high school diploma (%)	Share of farms that received technical assistance (%)	Share of farms that obtained rural credit (%)	Share of farms that are cooperative members (%)	Share of farms under contract farming (%)
Brasil	**5,175,636**	**333,680,037**	**64**	**163,986**	**35,350**	**24.5%**	**10.1%**	**24.0%**	**17.8%**	**10.6%**	**1.7%**
North	475,778	55,535,764	117	9,142	22,138	18.9%	6.4%	15.9%	8.9%	3.0%	0.9%
Northeast	2,454,060	76,074,411	31	29,219	13,128	41.0%	5.6%	8.6%	13.3%	1.8%	0.7%
Southeast	922,097	54,937,773	60	52,879	66,278	10.7%	18.0%	33.0%	15.2%	15.9%	1.3%
Minas Gerais	551,621	33,083,509	60	20,794	42,112	14.6%	13.3%	27.6%	16.8%	15.1%	1.1%
Espírito Santo	84,361	2,839,854	34	2,535	33,853	7.4%	14.8%	27.8%	16.8%	12.5%	0.8%
Rio de Janeiro	58,493	2,059,462	35	1,363	28,238	7.7%	18.7%	31.7%	5.8%	17.3%	0.8%
São Paulo	227,622	16,954,949	74	28,188	155,799	3.2%	30.3%	52.4%	13.3%	18.8%	2.1%
South	1,006,203	41,781,003	42	43,926	46,807	4.7%	12.9%	54.8%	36.5%	30.8%	4.9%
Center-west	317,498	105,351,087	332	28,820	109,150	8.4%	19.5%	32.0%	14.0%	10.9%	1.5%

Source: IBGE (2006).

Table 4.2 Agricultural land use in São Paulo state

	1995–1996		2007–2008	
	Number of agricultural production units	Area (hectares)	Number of agricultural production units	Area (hectares)
Perennial crops	84,382	1,332,694	83,971	1,225,035
Oranges	35,883	865,802	20,720	741,316
Coffee	28,399	229,090	23,737	214,790
Temporary crops	188,031	4,619,155	168,104	6,737,699
Sugarcane	70,111	2,886,313	99,799	5,497,139
Corn	84,910	1,235,906	51,694	667,685
Soybeans	9,441	714,207	7,816	396,427
Beans	18,056	162,208	10,290	104,154
Pastures	217,791	10,274,801	234,148	8,072,849
Planted trees	39,404	812,183	43,906	1,023,158
Natural Vegetation	108,881	1,954,151	155,211	2,432,912
Other uses	–	1,006,500	–	1,012,453
Total	**277,124**	**19,999,484**	**324,601**	**20,504,107**

Source: Department of Agriculture of the State of São Paulo, LUPA project, available from: www.cati. sp.gov.br/projetolupa.

today accounts for 27% of agricultural land use and 50% of the value of agricultural production in the state (Table 4.2). In addition to pastures, other relevant crops include oranges, coffee, and row crops. The state also has about 1 million hectares of planted trees, primarily *Eucalyptus* and *Pinus*, which are used in the production of wood panels, paper, pulp, and biomass for energy purposes.

A distinguishing feature of the main agriculture-based value chains in São Paulo — sugarcane, oranges, and planted trees — is that they are organized by large, vertically integrated agribusinesses. These three sectors are among the most competitive in the world based on high levels of productivity at the farm and the processing levels and vertically integrated supply chains from farms to customers in domestic and international markets. In addition, industry organizations provide public goods to industry participants helping them increase productivity and competitiveness. This chapter analyzes the organization and competitiveness of the sugarcane and orange juice sectors.

4.2 SUGARCANE

The Brazilian sugarcane industry is comprised of about 70,000 sugarcane producers, 400 processing units (sugarcane mills and distilleries) controlled by 160 economic groups, and 1.2 million workers. The gross domestic product of the sector was estimated at US$43 billion in 2014 (Neves and Trombim, 2014). Sugarcane production is carried out in 9.5 million hectares of which 5.5 million hectares are in the state of São Paulo. The Brazilian government introduced an agro-ecological zoning policy in 2009 to delimit areas where sugarcane (and other crops) may be produced. According to this policy, the permitted land area to grow sugarcane cannot exceed 65 million hectares or about 7.5% of the Brazilian territory. This law also banned agricultural production in sensitive biomes, such as the Amazon rainforest and the Pantanal wetlands, and limited agricultural expansion into native vegetation in the cerrado region.

Sugarcane production is clustered around two main regions: along the northeastern coast and in southeastern states around São Paulo (Figure 4.1). Although the industry was first established in northeastern Brazil, the region represents less than 15% of total industry output, with the remaining 85% produced in the southeast. About 60% of total sugarcane production occurs in the state of São Paulo, which has the highest productivity levels in the country. We examine below how the state became the center of the sugarcane industry and reached high levels of productivity and international competitiveness.

The industry output is impressive: 658 million metric tons of sugarcane were used as raw material to produce 38 million tons of sugar (equivalent to 22% of world production), 27 billion liters of ethanol (30% of world production), and bioelectricity in the 2013/2014 crop year (Figure 4.2). All sugarcane mills and distilleries in Brazil are self-sufficient in electricity. Processing plants use sugarcane bagasse — the cellulosic residue left after sugarcane is crushed — to generate bioelectricity for self-consumption. The excess of this clean energy not used in the processing plants is sold to distribution grids, thereby contributing to the nation's electricity supply. Considering the use of ethanol as fuel and the bioelectricity produced from bagasse, sugarcane generated the equivalent of 16% of the Brazilian energy consumption in 2012. In what follows, we briefly describe the evolution of the sugarcane industry and how Brazil became a world leader in this sector.

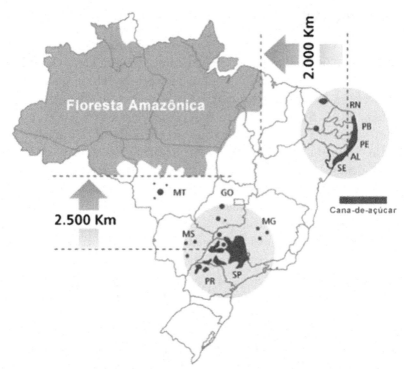

Figure 4.1 Sugarcane production areas in Brazil. *Source: UNICA (www.unica.com.br).*

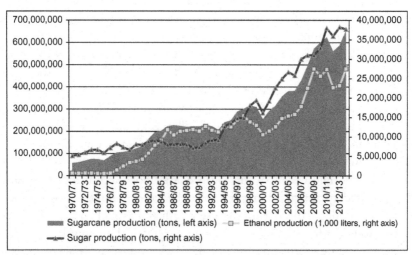

Figure 4.2 Output growth in the Brazilian sugarcane industry (1970–2014). *Source: Elaborated by the author based on data available in the Statistical Yearbook of Agroenergy, Ministry of Agriculture, Livestock and Food Supply (2013).*

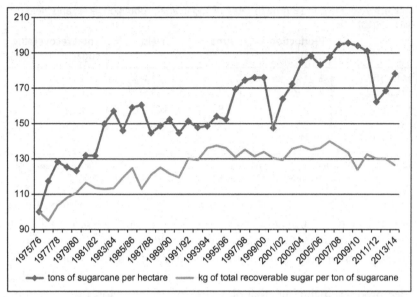

Figure 4.3 Productivity gains in the Brazilian sugarcane industry (1975–2014). (Productivity index (1975/76 = 100)). *Source: Elaborated by the author based on data available in the Statistical Yearbook of Agroenergy, Ministry of Agriculture, Livestock and Food Supply (2013).*

4.2.1 Industry Evolution

Sugarcane is an integral part of Brazil's social, political, and economic history.[1] One of the first decisions Portuguese conquerors made after landing in the southern coast of Bahia in 1500 was to introduce sugarcane brought from India and East Asia. Sugarcane producers were given large tracts of land by the Portuguese crown and used slave labor to produce sugar – the country's first export crop. Sugar was produced in large, vertically integrated plantations located along the northeastern coast. Between 1530 and 1650, sugar was the country's most important economic activity.

It was not until 1975 that the sugarcane industry started to become less dependent on sugar exports, when it received massive investments in science and technology from both private and public sources. These investments led to impressive production growth based on both sugarcane area expansion and productivity gains at the farm and processing levels (Figure 4.3; Table 4.3). The volume of total recoverable sugar (TRS),[2] which is the raw product used to produce sugar and ethanol, increased

Table 4.3 Average annual growth rates of sugarcane production, area, and yield (1961–2010)

	Production	Area	Yield	Total recoverable sugar
1961–1975	3.21%	2.45%	0.76%	−0.47%
1975–1985	9.70%	6.78%	2.92%	2.11%
1985–2000	2.38%	1.54%	0.84%	0.86%
2001–2014	7.47%	6.78%	0.69%	−0.35%
1961–2014	4.95%	3.68%	1.27%	0.60%

Source: Elaborated by the author based on data available in the Statistical Yearbook of Agroenergy, Ministry of Agriculture, Livestock and Food Supply (2013) and FAOSTAT.

from 4,000 kg per hectare in 1970 to 10,000 in 2010. Production of ethanol per hectare of sugarcane increased from less than 500 liters in 1970 to 5,000 liters in 2010. As a result, inflation-adjusted ethanol prices in the domestic market now cost less than one-third of what they did in 1975. The industry started to convert sugarcane into a diverse range of value-added products including ethanol, bioelectricity, and bioplastics.

The first defining moment in this process occurred in 1975 when the Brazilian government enacted the National Alcohol Program – known as *Pro-alcohol* – to reduce the country's dependence on foreign oil. The major pillars of *Pro-alcohol* included investment incentives for the construction of ethanol distilleries attached to existing sugar mills; a 5% mandatory ethanol blend in all gasoline sold in the country, which was gradually increased to the current level of 27%; and incentives to the production of pure-ethanol-powered vehicles. By means of price controls and subsidies, the government guaranteed ethanol producers a minimum price above production costs and set ethanol prices lower than gasoline at the pump for consumers. Petrobras, the state-owned oil company, received a mandate to distribute ethanol across the country and gas stations in towns with population above 1,500 were required to install ethanol pumps. With these measures, the government played a crucial role in the emergence of the bioethanol industry in Brazil. In 1975, less than 15% of the sugarcane production in the country was used to produce ethanol. By 1985, it had increased to more than 70%. Many experts agree that the industry would not have been able to take off without the big push of the *Pro-alcohol* period.[3]

In addition to the incentives provided under the *Pro-alcohol* program, the industry was heavily regulated until the beginning of the 1990s. Federal law 4870 enacted under a military dictatorship in 1965 defined the "rules of the game" from sugarcane fields to sugar and ethanol production, distribution, and exports. Prices were set at each stage along the value chain and each mill and distillery was allocated production and export quotas. Processors were also required to acquire at least 40% of their sugarcane supply from independent growers to limit the extent of vertical integration. The Sugar and Ethanol Institute (IAA) was the federal agency in charge of regulating the industry. This institutional setting tied the hands of the private sector and restricted competition. As a result, the industry mindset was production-driven. Industry participants also engaged constantly in lobbying activities as profit margins and industry growth were decided at the corridors of the IAA in Brasilia, the nation's capital.

Democracy was restored in the late 1980s and a new constitution was enacted that significantly altered the role of the state in the economy. Starting in the early 1990s the economy was liberalized, Brazil joined the Mercosur trade block, and the Real Plan was adopted to control inflation. The sugarcane industry embarked on a gradual process of deregulation starting with the dismantling of the IAA in 1990. A federal law enacted in 1994 discontinued all price and quantity controls and also liberalized sugar exports. In 1997 the ethanol domestic price control was extinguished. During this transition period, industry participants became increasingly driven by competitiveness and profitability. Following the end of *Pro-alcohol* in 1985 and industry deregulation in the 1990s, sugarcane production leveled off and the industry started to consolidate. One of the industry consolidators was Cosan, which acquired six processing plants between 1986 and 2000 to become the industry leader. But still the overwhelming majority of sugarcane processors were single-plant, family-owned firms.

The industry entered a new growth cycle after favorable market developments in Brazil and abroad, beginning with the introduction of flex fuel vehicles (FFVs) in 2003. FFV technology allowed consumers to fuel their cars with gasoline, ethanol, or any mixture of both. That is, fuel choice could be made at fueling stations, reducing risks for car owners and allowing the market to self-regulate based on the relative prices of each fuel. FFV technology was very popular among consumers and, by the end of the 2000s, over 90% of all new light vehicles sold in Brazil were FFVs. The FFV fleet reached 18 million vehicles in 2012 or approximately 60% of the light vehicle fleet in the country. Domestic ethanol demand increased in a similar pace to FFV sales, with ethanol use surpassing total gasoline demand in 2008. Domestic ethanol consumption increased from 10 to 23 billion liters between 2001 and 2010, including anhydrous ethanol blended in gasoline and hydrous ethanol (E-100).

Another breakthrough was the introduction of the renewable fuels standard (RFS) in 2005 followed by the 2007 Energy Independence and Security Act (RFS 2) that significantly increased the mandate for renewable fuel use in the United States. The RFS legislation set an ambitious target of 136 billion liters of renewable fuels by 2022. Other countries followed the US initiative to create a market for renewable fuels, including the Renewable Energy Directive of the European Union. Ethanol exports from Brazil increased from less than 100 million liters in 2001 to 3.2 billion liters (60% of the world's total) in 2009. As a result of consumption growth in developing countries and the export decline by previously large suppliers, particularly the EU and Cuba, Brazilian sugar exports also surged from 11 million tons in 2001 to 28 million tons in 2010, which represented 50% of the world's market.

With increased demand for sugar and ethanol, the industry entered a new phase of rapid growth and structural change in the 2000s. Sugar and ethanol processors engaged in joint ventures to make the necessary investments in logistics infrastructure and thereby take advantage of scale economies in distribution, exports, and risk management. The industry accelerated its consolidation process with several mergers and acquisitions (M&A). According to industry sources, 99 M&A transactions involving sugarcane processors occurred between 2000 and 2009. Family-owned processors began to hire professional managers and adopt corporate governance best practices. Some domestic firms converted to publicly traded corporations to access outside sources of capital with IPOs in Brazil and New York. Copersucar − a cooperative of sugarcane processors in São Paulo − adopted a hybrid ownership structure model to generate more capital for investments and growth. More than 100 new greenfield mills and distilleries were built across the country in non-traditional areas in São Paulo and adjoining states. Foreign players − including Sucden, Tereos, Dreyfus, Bunge, Cargill, ADM, Noble Group, Adecoagro, and Shree Renuka Sugars − and oil companies Shell, BP, and Petrobras entered the industry buying existing plants and building new ones. With leadership from UNICA, an association of sugarcane processors, the industry started to adopt more sustainable practices to decrease environmental and social impacts.

This structural change process significantly altered industry structure and performance. Before the start of the consolidation process, the great majority of the processors were family-owned, single-plant operators. The end of subsidies with the winding-up of the *Pro-alcohol* program and

Table 4.4 Size and ownership structure of largest sugarcane processors in Brazil (2009—2010)

	Processed sugarcane (tons)	Ownership structure
Copersucar	68,322,123	Cooperative
Cosan	52,781,685	Publicly traded corporation
LDC (Louis Dreyfus)	19,388,223	Multinational
Tereos	13,652,029	Multinational
São Martinho	12,923,436	Publicly traded corporation
Bunge	9,285,292	Multinational
São João Araras	7,371,057	Family-owned
Cerradinho	6,588,721	Family-owned
Shree Renuka Sugar Ltd	6,582,275	Multinational
Colombo	6,518,941	Family-owned
Bazan	6,110,957	Family-owned
Grupo Toniello	4,728,588	Family-owned
Luiz Cera Ometto	3,606,616	Family-owned
ETH Odebrecht	2,832,469	Publicly traded corporation
Other 28 firms	53,580,386	—
Total	**274,272,798**	—

Note: This list only includes sugarcane processors that are members of UNICA.
Source: UNICA.

the subsequent industry liberalization provided strong incentives for processors to increase efficiency or exit the industry. Family businesses started to professionalize their management and pursue economies of scale either by increasing plant size or by acquiring other plants. In general, plants controlled by low-ability entrepreneurs that were attracted to the industry with the subsidies of the *Pro-alcohol* phase were acquired by more successful ones. Agricultural best practices were transferred to acquired plants with positive effects on their operating performance.[4]

Another major structural change was internationalization. Until 2000 only domestic firms operated in the sugarcane industry. The first multinational companies to enter the industry were French sugar traders Dreyfus, Sucden, and Tereos. By the end of the decade, the sugarcane industry was comprised of firms with different ownership structures, including cooperatives, publicly traded corporations, multinational companies, and family-owned businesses (Table 4.4). Industry sources estimated that multinational players controlled 32% of the industry capacity in 2011. In addition to capital, multinational firms brought management capabilities that raised industry standards to higher levels. The entry of new players in the last 15 years also accelerated industry consolidation. The five largest industry groups, all of them controlled by domestic owners, had a combined sugarcane-processing capacity of 77 million tons or 19% of total production in 2006. The combined capacity of the five largest groups increased to 173 million tons or 26% of total production in 2011. Despite this consolidation wave, no producer accounts for more than 10% of domestic production in terms of crushed sugarcane and the industry structure is still fragmented, with low levels of economic concentration.

After the 2008 global financial crisis, the industry entered a turbulent period. Sugar and ethanol exports decreased and many of the growth opportunities that fostered massive capital investments did not materialize. Sugarcane processors that had invested heavily to build new plants or to acquire existing ones were highly leveraged. In addition, the government decided to reduce the tax on gasoline consumption and control gasoline prices at the pump to control inflation, which squeezed ethanol margins and further strained the balance sheet of processors. The financial constraints of processors further exacerbated the crisis as necessary investments to renovate the sugarcane fields were forgone, with negative effects on productivity and operating performance. To make things worse, the industry suffered the negative effects of three consecutive years of bad weather conditions that led to crop failures. In 2014, of the existing 400 processing plants in the country, 80 were closed and 67 were in bankruptcy proceedings.

In what follows, we take a closer look at some case studies highlighting the structural changes that occurred in the sugarcane industry in the last decades. We start with the case of UNICA to examine the role of a producers' association in industry leadership and development.

4.2.2 The Brazilian Sugarcane Industry Association (UNICA)

The history of UNICA started in 1932 with the formation of the Sugarcane Millers Association by processors in the state of São Paulo. Between 1932 and 1990, the association office was housed at the Copersucar headquarters along with the sugar and ethanol processors' unions. The presidents of processing firms — the majority of which were family-owned businesses — took turns in managing the association. With the enactment of *Pro-alcohol* in 1975 many processors decided to form competing industry associations. UNICA was formed in 1997 as a union of these rival associations to provide a unified industry position in a deregulated market environment.

In 2000 UNICA members decided to hire Eduardo Pereira de Carvalho as its first professional president and CEO. With extensive industry experience, Antonio de Padua Rodrigues was hired as the technical director to assist Eduardo. The board of directors maintained responsibility for setting the policies and providing strategic direction, but execution was delegated to a professional staff with considerable autonomy. Eduardo changed the organizational structure of UNICA and introduced objectives, goals, and performance measures for the management staff. Eduardo led UNICA until 2007 with a focus on increasing industry competitiveness in a deregulated market environment. His major accomplishments were to consolidate UNICA as the unified industry voice and to introduce professional management to UNICA, which was rare among industry associations in Brazil at that time.

By the late 2000s the industry dynamics had changed again with significant growth, entry of new players, and structural change. But the Brazilian sugarcane industry started to become the target of attacks and accusations. Opponents argued that sugarcane ethanol was a cost-effective alternative to gasoline, but that it destroyed native forests, it employed labor under slave-like conditions, and it was responsible for food price inflation. The industry was not ready to face these criticisms and adopted a distant, passive approach. This started to change in 2007 when Marcos Jank was hired as the new president and CEO with a mandate to develop a sustainability agenda and to better communicate with outside stakeholders. In what follows we examine how UNICA helped transform the industry during Marcos's tenure between 2007 and 2012.

4.2.2.1 Governance and Organizational Structure

UNICA members are sugarcane processors located in São Paulo and adjoining states. In 2009, UNICA counted 41 members that owned 123 processing plants and crushed 274 million tons or about 50% of the Brazilian sugarcane crop. Membership is voluntary and open but applications of new members have to be approved by the board of directors. Membership fees and voting rights in the association are set in proportion to sugarcane crushing volume. As a result, the largest processors contribute more to UNICA's budget but also control more board seats.

The UNICA governance structure is based on a three-tiered model: the board of directors, three committees, and the executive team. The board is responsible for making strategic decisions and setting policy. It is comprised of 24 elected seats in addition to the president-CEO. Each director is elected for a 3-year term with no term limits. Board meetings occur every Tuesday afternoon at the UNICA office in São Paulo. The last board meeting of each month, when UNICA staff brief members about current industry affairs, is open to all members. "These plenary monthly meetings are very important to our members as it is also an opportunity for them to interact socially. Our association has the culture of a club," according to Eduardo Leão de Sousa, the executive director. The governance structure of UNICA also includes a fiscal board and three technical committees. A general assembly of members is held once a year to approve financial statements and the budget and to conduct the election of board directors.

The execution of the strategic and action plans laid out by the board is the responsibility of the professional staff. UNICA's organizational structure includes the president-CEO and three directorships – executive, technical, and communications. The CEO and the three directors form the Executive Committee. The staff includes full-time employees, executives, and consultants bringing a diverse set of skills and experience to UNICA. The professional team is also in charge of coordinating several technical commissions. These commissions are formed on a non-permanent basis to discuss relevant issues to the industry with the participation of members, non-members, and industry specialists.

4.2.2.2 Sustainability Initiatives

Between 2007 and 2012, the UNICA team worked on several international and domestic fronts to introduce industry-wide sustainability initiatives. These initiatives included engagements with foreign governments, multistakeholder organizations, NGOs, labor unions, and several federal and state agencies in Brazil. First, UNICA interacted with foreign government officials and legislators to influence the development of policies and regulations concerning renewable sources of energy such as the Renewable Fuel Standard (RFS) and California's Low Carbon Fuel Standard (LCFS) in the US and the EU Renewable Energy Directive. These policy processes were critical to the industry as they had the potential to open or close markets for sugarcane ethanol. UNICA established foreign offices in Washington and Brussels to coordinate more closely its lobbying efforts and influence the policy debate in a timely fashion. UNICA also participated in discussion groups and roundtables, including the Roundtable of Sustainable Biofuels and the Better Sugarcane Initiative (Bonsucro). These multistakeholder initiatives (MSIs) intended to regulate business behavior and promote sustainable business practices with the development of certification processes. They were formed by a broad range of participants such as NGOs, civil society organizations, trade unions, and multinational corporations.[5] UNICA decided to participate in MSIs to represent the interests of producers from a developing country perspective. Geraldine Kutas, the head of International Affairs, argued that the main challenge in these roundtables was "to close the gap between the sustainability demands of consumers, processors and retailers in the developed world and the realities faced by commodity producers in developing countries. In addition, nobody wants to bear the increased costs associated with sustainability certification of a commodity — such as sugar and ethanol — and the producer always ends up bearing these costs." Despite these challenges, she believed MSIs were very important to open direct channels of communication and build trust between stakeholders representing different perspectives.

Eduardo Leão de Sousa was in charge of the domestic front, including engagements with Brazilian government officials, policymakers, consumers, labor unions, and NGOs leading to certification of sustainable practices. Eduardo believed that achieving sustainability should involve "a two-way communication process as information must flow upstream to sugarcane producers and they must be ready to respond to the demands of customers and end consumers." Examples of certification of sustainable practices involving the sugarcane industry included the Green Protocol, the National Labor Commitment, and the *RenovAção* project.

The Green Protocol was signed in June 2007 by UNICA and the São Paulo governor and secretaries of agriculture and the environment to promote sustainable environmental practices in sugarcane production and processing in the state. The protocol established a series of guidelines to be voluntarily followed by processors seeking eligibility for the Certificate of Environmental Compliance. These guidelines consisted of practices related to soil and water conservation, riverside forest protection, greenhouse gas emission reduction, and responsible pesticide use. Despite the breadth of the protocol, the most important directive was the introduction of sugarcane harvest mechanization in place of the traditional practice of sugarcane burning that allowed workers to manually harvest the fields. Prior state legislation required sugarcane burning to be eliminated by 2021. The Green Protocol agreement advanced this deadline to 2014. More than 170 sugarcane mills and 29 producer associations have voluntarily adopted the protocol since 2007. In 2013, 89% of the sugarcane harvested area was mechanized, compared to 26% in 2006.

In June 2009 the National Commitment for the Improvement of Labor Conditions in Sugarcane Production was launched by the Brazilian federal government, UNICA, the Federation of Rural Workers in the State of São Paulo (FERAESP), and the National Confederation of Workers in Agriculture (CONTAG). The main purpose of the National Labor Commitment (NLC) was to incentivize and recognize the adoption of best labor practices in the sugarcane industry. The Brazilian sugarcane industry employed approximately 1.2 million workers in both the farm production and processing sectors. Although the industry had made significant progress in improving work conditions, labor issues still persisted even among some large processors. Processors that voluntarily committed to the program seeking to receive the Conformity Certificate had to follow 30 guidelines set forth by the terms of the agreement. These guidelines included practices that were stricter than the legal obligations of federal labor laws. They addressed issues related to safety, health, and general working and hiring conditions of workers engaged in manual operations in sugarcane fields. In 2014, 180 sugarcane mills had been certified by independent auditors and 75 more were in the process of being certified.

The RenovAção project was a training program created by UNICA in partnership with the state rural workers' union (FERAESP) and the Solidaridad Foundation. The project also received support from the Inter-American Development Bank (IDB) and funding from industry partners Syngenta, Case IH, Iveco, and FMC. The initiative was launched in 2009 as a response to the fast mechanization of sugarcane planting and harvesting triggered by growing environmental and social concerns. The phasing-out of pre-harvest burning and manual harvest suggested that a great number of workers employed as sugarcane cutters would eventually lose their jobs. The objective of the RenovAção project was to train these workers in six sugarcane production areas in the state of São Paulo. The training program was divided into two major components: courses to reposition cane cutters within the sugarcane industry (e.g., as mechanical harvester operators, mechanics, truck drivers, electricians, etc.) and courses to reposition displaced cane cutters in other sectors of the local economy (e.g., construction, pulp and paper mills, and horticulture). Course offerings targeted local opportunities and specific labor demands in each affected community. Between 2009 and 2012, the project trained more than 21,000 rural workers with 78% of them being able to find another job.

In addition to providing industry leadership and representing members in the negotiation and development of certification processes, UNICA coordinated the development of corporate social responsibility (CSR) efforts at the processor level. Since it had signed agreements such as the Green Protocol and the NLC, UNICA needed to bring its members along to be able to deliver on its commitments. Because the adoption of sustainable practices by sugarcane processors was voluntary, UNICA's CSR team — led by Maria Luiza Barbosa — offered courses and leadership development programs for processors interested in adding CSR to their strategic initiatives.

In addition, the field staff used data obtained from processors to develop industry benchmarks for key social and environmental indicators. These indicators served as a management tool allowing processors to benchmark their sustainability performance against industry averages and best practices. Additionally, bankers, customers, and Brazilian society at large were increasingly demanding sustainable business practices. It was more and more difficult to get funding from major banks or do business with large customers if a processor did not follow sustainable practices. Maria Luiza believed that "when a sugarcane processor adopting sustainable practices signs a big supply contract with Coca-Cola or Nestlé, this is a major incentive for industry rivals to follow."

The combined CSR efforts and projects of UNICA members were compiled in the industry sustainability report. In 2009 UNICA became the first Brazilian industry association to publish a sustainability report based on the guidelines developed by the Global Reporting Initiative (GRI), an international organization based in the Netherlands. The GRI was created to provide levels of consistency and legitimacy to sustainability reports equivalent to those of financial reports. In its 2008–2009 sustainability report, UNICA described 618 CSR programs implemented by its members during that crop year. These programs in the areas of education, culture, health, quality of life, and the environment required annual investments of BRL 158 million and benefited 480,000 people living in the communities near sugarcane mills. UNICA's GRI-checked sustainability report also served as an important communication tool with outside stakeholders.

4.2.2.3 Communication Efforts

In the mid-2000s the Brazilian sugarcane industry was under considerable pressure from external stakeholders. The industry, however, had a culture of not responding to outside criticisms, leading to the perception that it lacked transparency. As the industry did not position itself relative to critics, misinformation and "myths" were widespread. Adhemar Altieri was recruited as UNICA's communications director in November 2007. He built a team of 12 professionals in charge of communication, marketing, public relations, and content management. His major goals were to provide information about the sugarcane industry to all requests, to correct erroneous information published or broadcast about the industry, and to collect and organize information about major industry advances that had been systematically overlooked by the media and other outside stakeholders. To support this proactive communications strategy, UNICA invested in the internal production of information for outside stakeholders and marketing campaigns with a message focused on the benefits of ethanol as a green and sustainable source of energy.

4.2.2.4 A Brief Update

For two consecutive years (2012–2013), UNICA's budget was frozen and in 2014 the board decided to reduce members' contributions by half due to the industry crisis. UNICA had to downsize and refocus its priorities. The communication department was significantly reduced to one person and the team responsible for working directly with the sugarcane mills in sustainability benchmarking was let go. The current strategy prioritizes engagement with the federal government to increase industry competitiveness and seek solutions to ameliorate the difficult financial situation of processors and the industry as a whole. UNICA hired Elizabeth Farina, former chair of CADE (the Brazilian anti-trust commission), and an expert in industrial organization and regulation, as the new CEO in 2012. Eduardo Leão, the executive director, moved his office to Brasilia to focus on government relations.

Despite this strategic reorganization, UNICA's efforts in sustainability and communication are paying off. More than 40 processing plants are certified by Bonsucro with sustainable sugarcane production practices. The industry has been able to improve labor conditions and also to deliver on its promise to eliminate sugarcane burning in São Paulo. About 70 sugarcane processors adopted sustainability practices and reporting following GRI guidelines. Perhaps more importantly, the industry now operates under more stringent rules and has achieved a better standing with its external stakeholders.

4.2.3 The Relationship Between Sugarcane Processors and Growers

Historically, the relationship between sugarcane growers and processors has been one of conflict. So much so that, since the 1940s, the government has implemented policies to set marketing margins along the value chain and to limit the extent of vertical integration by processors. Processors were required to procure at least 60% of their sugarcane supplies from independent growers. Between 1950 and 1985, the relationship between sugarcane production by processors (vertical integration) and production by independent growers was stable at around 50% each. After the end of the *Pro-alcohol* program in 1985 processors increased the share of vertical integration to guarantee the supply of raw material and to decrease transaction costs with growers. Vertical integration also allowed processors to have more control over the agricultural production process and to adopt technologies and practices to increase sugarcane productivity. Between 1985 and 2000, the share of vertical integration increased to almost 70% of the sugarcane processed in the country. Since then, it has decreased to 60% as a result of entry of new players and industry growth into new production areas.

As we discussed above, the efficiency of sugarcane production is vital to the operating performance of a sugarcane processor. It represents about 70% of the total production cost of sugar and ethanol. The more sugar a field can produce, which depends on both sugarcane productivity per hectare and the amount of sugar in the cane, the more output (sugar and ethanol) and the lower the per-unit costs of production. Increased sugarcane productivity depends crucially on plant genetics, agronomic practices and, of course, Mother Nature. In addition, sugarcane is a low-value, high-volume product, which means that sugarcane fields need to be located close to the processing plant to minimize transportation costs. In general, it is not economically feasible for a processing plant to source sugarcane beyond a radius of about 50 km around it. Finally, sugarcane is a perishable product that cannot be stored after harvest. The longer the time between the sugarcane harvest and processing, the lower the quality of sugar in the cane, which hurts the operational efficiency of the processing plant. This means that sugarcane planting, harvesting, and transportation operations must be carefully coordinated with the production schedule of the processing plant, which in general runs at capacity 24-7 for 9 months of the year. In other words, there are strong economic incentives for processors to vertically integrate upstream into sugarcane production or at least control some key agricultural decisions and the planning of harvesting and transportation to the mill.[6] This is achieved with a multiple sourcing strategy combining vertical integration with long-term production contracts with growers.

Following industry deregulation in the early 1990s, sugarcane, sugar, and ethanol prices were no longer set by the government but by the market forces of supply and demand. A group of Brazilian sugarcane producers that were concerned with how sugarcane would be priced in the free market decided to examine pricing arrangements between farmers and processors in other countries. A committee made up of representatives of sugarcane producers and processors was put together to develop a transparent model to value and price the sugarcane supplied by independent growers to processors based on total recoverable sugar (TRS). TRS measures the quantity of sugars extracted from the sugarcane that are used as raw material in the production of sugar and ethanol in the processing plants.

This work culminated in the creation of the Council of Sugarcane, Sugar, and Ethanol Producers (Consecana) in 1999. Since then, Consecana has been responsible for managing the relationship between sugarcane growers and processors and setting the base price of TRS. It operates similarly to marketing orders in the United States, but without government interference and based on voluntary adoption by industry participants. The Consecana board is elected by representatives of sugarcane growers (ORPLANA) and processors (UNICA).

Consecana began operating in the 1998–1999 crop year, with 85% of sugarcane production being marketed based on its pricing formula. The sugarcane pricing formula is based on TRS and sugar and ethanol prices in the domestic and international markets. The Consecana board sets the sugarcane base price every year following technical and economic recommendations from a technical committee. The Consecana pricing formula has been effective in increasing market transparency, decreasing transaction costs and conflicts between growers and processors, and providing the basis for transfer pricing between the agricultural and processing divisions of vertically integrated firms.

4.2.4 The Industry Consolidators: Copersucar and Cosan

The Sugarcane Producers Cooperative (Copersucar) was formed in 1959 by ten sugarcane processors and two other regional cooperatives in

São Paulo. Copersucar was founded to market the sugar and ethanol production of its members but it also played a leading role in the evolution of the sugarcane industry. It housed the Sugarcane Millers Association in its offices until 1990 and helped shape the policy and regulatory environment affecting the industry, especially during the *Pro-alcohol* years. Copersucar is a member of UNICA with considerable influence. Given its large production volume, it appoints seven of the 25 board members. Copersucar also played a major role in the technological development of the industry. It invested heavily in agricultural R&D through the Copersucar Technological Center (CTC), which was established in 1969 to develop new sugarcane varieties adapted to São Paulo production conditions and other technologies that fostered productivity gains in the state (see Section 4.2.5).

In its first 40 years Copersucar grew horizontally with production growth of members and entry of new members, but it also diversified its operations. It had the leading sugar brand name in the domestic market (*União*) and a relevant coffee roasting operation. By the end of the 1990s, Copersucar was the largest cooperative in the country, with more than 12,000 employees. However, the structural changes of the 2000s forced the cooperative to reinvent itself. With industry consolidation and the entry of multinational players, Copersucar needed to restructure not to lose market relevance. After a long soul-searching process, the Copersucar board decided to refocus the cooperative on its core business of sugar and ethanol marketing, trading, and logistics.

In 2000 Copersucar sold its coffee assets to Sara Lee and in 2004 it divested of its sugar retail operation and the *União* brand. Luís Pogetti, the current chairman of Copersucar, shared with me in a personal interview that "divesting of the União brand was my first big challenge when I joined Copersucar. Our members had an emotional bond with the brand and there was strong resistance to its sale. But in the end it proved to be the right decision." Copersucar also spun off its R&D department (CTC) and allowed other processors to become shareholders. After the reorganization, the number of employees decreased from 12,000 to 160.

However, asset divestitures were not enough to generate the necessary capital to invest in growing its core business. To be a relevant player in sugar and ethanol required billions of dollars of investments in logistics, operational improvements, and scaling-up. The goal was to become the number-one originator and marketer of sugar and ethanol in the world with a market share of 30% of the Brazilian sugarcane production by 2018.

"That's when we realized that our traditional cooperative structure had reached its limit. In the cooperative structure, members benefit as suppliers of sugar and ethanol, not as investors. They do not earn a return on investment and there is no mechanism for equity capital valuation and transferability. We needed growth capital to execute our strategic plan, but our members did not have incentives to invest," according to Mr. Pogetti.

Following a 3-year process, Copersucar members decided to change its ownership structure and a hybrid cooperative—corporate model was adopted in 2008.[7] In this new structure, the sugarcane processors continued to be members of the Copersucar cooperative and market all their sugar and ethanol production through the cooperative. The downstream assets related to sugar and ethanol logistics, trading, and marketing were transferred to a corporate entity (Copersucar S.A.) owned and controlled by the same owner-members of the cooperative via a holding structure (Produpar). Both the cooperative and the corporate entities shared the same team of professional managers. In addition, modern corporate governance practices were adopted, including separating the roles of chairman and CEO, the appointment of two independent directors to the board, and the adoption of a corporate code of conduct and a risk management policy.

Since the 2008 restructure Copersucar became the largest producer and exporter of sugar and ethanol in the world. Between 2009 and 2014, the number of sugarcane mills that are members of the cooperative and shareholders of Produpar increased from 22 to 43; sugarcane crushing increased from 46 million to 135 million tons;[8] and revenues soared from BRL 4 to BRL 23 billion. In 2014 Copersucar marketed 8.6 million tons of sugar in both domestic and international markets. This volume represented 25% of total sugar production in south-central Brazil and 11% of the global sugar export trade. In the same year, ethanol production was 4.9 billion liters or 19% of the south-central production. In 2012 Copersucar acquired a controlling stake in Eco-Energy, a marketer of corn ethanol in the US. The combined marketing volumes of Copersucar and Eco-Energy had a market share of 12% in global ethanol trade. In 2014 Copersucar announced a joint venture with Cargill in sugar origination and trading, further expanding its global reach and industry leadership.

To market such a huge volume of sugar and ethanol in Brazil and international markets, investments in logistics were projected to add up to BRL 2 billion by 2015, including a sugar export terminal in the port of Santos, an ethanol terminal in Paulínea, SP (a major petrochemical hub in the country), and a 20% share in Logum S.A., an ethanol pipeline linking sugarcane production regions to distribution terminals in São Paulo and Rio de Janeiro. Copersucar also signed long-term contracts with railroads to transport sugar by rail to the port instead of trucks. The logistics network built by Copersucar serves its own needs but also generates income by providing services to third parties.

Copersucar has a unique business model. The 43 sugarcane mills are owned and operated by 24 entrepreneurs. They specialize in what they do best – growing sugarcane, managing relationships with independent growers, and producing sugar and ethanol. A vertically integrated sugarcane growing and processing operation is highly complex and prone to substantial risks from Mother Nature. It is also a capital-intensive, commodity business that requires a focus on operational efficiency and local knowledge to coordinate agricultural, transportation, and processing operations. Because of these operational complexities, the potential for synergies in multiple plant operations is limited. Mr. Pogetti confided that "I would not be able to manage 43 processing plants and thousands of suppliers from my office in the city of São Paulo." These processing plants supply all their sugar and ethanol production to Copersucar under a long-term contractual agreement. Copersucar also originates production from non-member processors on a competitive basis. Copersucar pays market prices for both members and non-members. In doing so, Copersucar focuses on gaining economies of scale and market relevance in sugar and ethanol logistics, marketing, and trading. In Mr. Pogetti's words, "we consolidated the supply, not the assets."

One of the members of Copersucar is *Usina São Manoel* (USM) owned by the Dinucci family. The processing plant was built in 1949 after the previous owners converted the farm from coffee to sugarcane production. The Dinucci family, which had a construction business, bought the farm and the plant in 1955. They joined Copersucar in 1959 to gain more bargaining power in marketing sugar. After the passing of his father, Carlos Dinucci became the chairman and CEO of USM in 1980. A civil engineer by training, Mr. Dinucci did not know much about agriculture and sugarcane processing. "I became the CEO of USM by necessity."

The first decade as CEO was challenging for him. In our personal interview, Mr. Dinucci was candid enough to admit that "our performance until 1990 was mediocre. Our agricultural productivity was 55 tons of sugarcane per hectare, which was ridiculous." He then decided to follow more closely the technical recommendations of CTC staff to adopt better technologies and agronomic practices. In 1992 he took a 3-month course on total quality management in Rochester, NY, and decided to apply the management principles developed by W.E. Deming in USM with the help of a consultant. "Since then we have introduced the total quality management philosophy in our business and worked tirelessly to instill a continuous improvement culture among our collaborators. It is relatively easy to adopt new technologies. The hardest part is to change the culture of the organization. Deming's management principles are simple and intuitive but they are very hard to implement without buy-in from collaborators. We have more than 3,200 collaborators and communication with them is the key."

In the 2000s, Mr. Dinucci followed UNICA's leadership and embraced the sustainability agenda. USM voluntarily adopted the Green Protocol and the National Labor Commitment, the plant is certified by Bonsucro and the ISO 22,000 protocol, ethanol production follows RFS2 and CARB standards, and its performance can be tracked with its annual sustainability report checked by GRI. "We benefited a lot from working with Maria Luiza Barbosa and her team at UNICA. They helped us organize our sustainability efforts, benchmark against best industry practices and communicate effectively with our stakeholders." Today USM is arguably one of the best-managed plants in Brazil. Between 2009 and 2014, which were turbulent times for the industry, USM generated above 40% cash flow (EBITDA) margin. Mr. Dinucci believes that "the industry crisis is a great opportunity for us to grow."

In 2014 USM processed 3 million tons of sugarcane in its plant and produced about 200,000 tons of sugar and 100 million liters of ethanol, which were marketed by Copersucar. Because the plant is certified by Bonsucro, USM received a premium over market prices. About two-thirds of the processed sugarcane is grown in land owned or leased by USM and the remaining third is supplied by some 100 independent growers under long-term production contracts. "The key to the success of a sugarcane processor is planning. You must be able to coordinate sugarcane production schedules with harvesting and transportation to the plant to maximize TRS available during the crop season and optimize plant utilization." Managing the relationship with growers is also critical because USM competes with two other plants in its region for sugarcane, one of them being the largest plant run by Cosan (called *Usina da Barra*).

An alternative consolidation model was pursued by Cosan. We explained above that Cosan was the first family-owned sugarcane processor to acquire other processing plants after the phasing-out of the

Pro-alcohol program in 1986. The original processing plant, called *Usina Costa Pinto*, was built in 1936 in Picacicaba, SP. After the acquisition of six other plants between 1986 and 2000, the firm adopted the name Cosan. Since then the growth of the company has been phenomenal. In 2005 Cosan S.A. listed in the Bovespa New Market valued at US$400 million. In 2007 Cosan Limited — the parent company of Cosan S.A. — was listed in the New York Stock Exchange (NYSE). Public listing required the company to adopt sound corporate governance practices and transparent reporting to shareholders.

With the money raised in its IPO, Cosan continued to expand horizontally with the acquisition of more plants. In 2008 the company entered the fuel distribution and lubricant production business with the acquisition of Exxon Mobil assets in Brazil. In the same year, it created a subsidiary called Radar to manage its land asset portfolio of 240,000 ha and another subsidiary called Rumo focused on sugar export logistics. In 2010 Cosan signed a joint venture agreement with Shell to create Raizen and in 2012 it acquired a 60% stake in Comgás, the largest distributor of natural gas in Brazil. In 2014 Rumo merged with Latin America Logistics (ALL), the largest railroad operator in Brazil. Following these acquisitions and joint ventures, Cosan became one of the largest conglomerates in the country with consolidated revenues of BRL 40 billion in 2014.

But let us take a closer look at its sugar and ethanol business, Raizen, a 50—50 joint venture between Cosan and Shell. Raizen is a BRL 57 billion company with 40,000 employees organized into two divisions: Raizen Energy (sugar and ethanol production) and Raizen Fuels (fuel distribution). Raizen Energy operated 24 plants with a combined crushing volume of 62 million tons of sugarcane and production of 2 billion liters of ethanol and 4 million tons of sugar in 2014. Ten of these plants are certified by Bonsucro. About half of the processed sugarcane is grown on their own (vertically integrated) plantations[9] and the other half is supplied by 3,000 independent growers under production contracts. Attached to these plants, Raizen has installed capacity of 940 MW, which is enough to supply energy from sugarcane bagasse to a city of 5.5 million people. Raizen Fuels has a distribution network of 5,200 Shell service stations and 60 distribution terminals across the country. Raizen also has a 20% equity stake in Logum S.A., the ethanol pipeline also owned by Copersucar and other companies, and in CTC.

While Copersucar lets its members specialize in sugarcane production and processing to focus on logistics and marketing, Raizen follows a completely integrated business model from sugarcane plantations to service stations. These are alternative business models developed by Brazilian entrepreneurs to grow and remain relevant players since the structural change process ushered in since the 1990s industry deregulation.

4.2.5 The Copersucar Technology Center (CTC)

The Copersucar Technology Center (CTC) was founded in 1969 with the objective of developing sugarcane varieties and agronomic practices adapted to the growing conditions of São Paulo soils and climate. At that time, only state research institutes such as IAC in São Paulo and public universities had research programs in sugarcane. The R&D budget was funded by voluntary contributions of Copersucar members based on their respective sugarcane processing volumes. With these contributions, CTC was able to build the largest sugarcane germplasm in the world and develop a successful breeding program with more than 80 sugarcane varieties. All industry participants benefited from the varieties developed and technical assistance services provided by the CTC staff, which followed an "open access" model. The CTC thus contributed to agricultural productivity gains of about 1.2% per year between 1970 and 2004. Producers in the state of São Paulo increased sugarcane production per hectare from an average of 65 tons in 1975 to 90 tons in 2009, consistently higher than their counterparts in other parts of the country. The production of ethanol increased from 2,500 to 7,000 liters per hectare.

However, productivity gains leveled off in the 2000s and then decreased after the 2009 crisis. There were several reasons behind this trend. Processors were highly leveraged and faced financial constraints, which forced them to reduce investments in the necessary renewal of sugarcane fields. Back-to-back seasons with bad weather further decreased productivity in 2010 and 2011. Adoption of harvest mechanization in São Paulo and sugarcane expansion to non-traditional areas with poorer soils and less-than-optimal growing conditions also contributed to lower productivity. Despite these external factors, R&D investments had not been enough to sustain high levels of productivity gains in the sugarcane industry. The major biotechnology and seed companies did not invest in sugarcane because the potential market is small. The planted area of sugarcane of about 25 million hectares around the world pales relative to the planted areas of wheat (225 million hectares), corn (160 million hectares), and soybeans (100 million hectares). In addition, the sugarcane plant is genetically more complex than other crops, which requires higher investments in R&D efforts. Perhaps more importantly, the CTC model was not conducive to generating enough funds for R&D investments because contributions by processors were voluntary and the knowledge generated was a public good. With such an unsustainable model, CTC lost market share to Ridesa, a consortium of 11 state and federal universities that coordinated their R&D and marketing efforts.

This situation needed to change. As part of its restructuring process, Copersucar decided to spin off CTC in 2004. It was established as a nonprofit organization and other processors were invited to join. The number of CTC members increased from the 31 Copersucar plants at that time to more than 180 plants in São Paulo and other states, representing 60% of the sugarcane processed in the center-south region. These plants were required to make contributions in proportion to their sugarcane processing volume to fund the CTC budget. Another major organizational change occurred in 2011, when CTC was reorganized as a corporation (S.A.) and received equity funding from BNDESPar, the private equity arm of the Brazilian Social and Economic Development Bank (BNDES). BNDESPar bought a 25% stake in CTC S.A. as a minority shareholder and required changes in its corporate governance practices. The board of CTC S.A. is now comprised of 11 directors — eight appointed by the processors, one appointed by BNDESPar, and two independent directors. The new CEO was hired from Monsanto and the management team was reorganized. Mr. Pogetti, who is also the chaiman of CTC, explained that "in the old structure, CTC lacked an economic rationale to be an efficient organization. In the new corporate structure, CTC has to perform. It has sufficient funding for its current R&D program and it will be able to charge royalty fees for the new varieties and technologies it creates. It will have to create value to its shareholders."

As a result of this reorganization, the R&D program of CTC was refocused and regionalized. The research focus shifted to disruptive technologies with the objective of doubling the industry innovation rate in the future. These disruptive technologies include molecular markers to accelerate the conventional plant breeding program, biotechnology to introduce insect and drought resistance traits and increase the amount of sugars in the cane (TRS), and second-generation ethanol. CTC is investing in new technologies and enzymes to be able to extract sugar from bagasse and leaves, with the potential to increase ethanol production per hectare by 50%. These new technologies are developed in collaboration with EMBRAPA, BASF, Novozymes, the University of São Paulo College of Agriculture (ESALQ), and other strategic partners.

The development of new varieties is now conducted in 12 regional centers located in the main production areas across Brazil. These regional centers develop new sugarcane varieties adapted to each region and local agronomic solutions with input from the processors. CTC has leased land in these regions to increase the area dedicated to multiplication of seedlings and speed up the adoption of new varieties in non-traditional areas. CTC has also invested in satellite imaging technology to support its precision agriculture program and provide better crop forecasts.

Another change implemented at CTC was the development of a commercial department. The commercial team of about 60 employees is in charge of pre-commercial testing of new seedlings in fields provided by processors, the marketing and pricing of new sugarcane varieties, and customer relationship including sales and technical assistance. Luiz Antonio Dias Paes, the manager in charge of the commercial team, explained that "there has been a major cultural change in CTC since the reorganization. We now have to create value to our shareholders and increase their competitiveness as sugarcane producers and processors with clearly defined performance measures. CTC researchers and the commercial team receive incentive compensation if we perform and deliver the expected results." Another major change is that customers must now pay royalty fees to have access to the new varieties. "We will charge the same royalty fee for all customers, irrespective of whether they are shareholders of CTC or not. This new business model will allow us to capture some of the value from the innovation created in our research fields and labs and generate sufficient resources to fund our research programs and ensure we are a viable organization in the future."

4.3 ORANGE JUICE

In addition to sugarcane, the state of São Paulo is also the center of the Brazilian orange juice industry. Brazil is the world's largest producer of oranges and orange juice and exports almost 80% of the frozen concentrated orange juice (FCOJ) consumed in the world. The industry is comprised of nine processors and 12,000 growers; it employs 230,000 workers and generates US$6.5 billion in gross annual income (Neves, 2010). Interestingly enough, world production of oranges destined for juice production is concentrated in two states: São Paulo (Brazil) and Florida

(United States). These two states combined produce 90% of the world's supply of orange juice. But while orange and juice production has decreased in Florida due to bad weather and the spread of diseases, the São Paulo industry has grown since the 1960s (Table 4.5).

Since its beginning the orange juice industry in São Paulo has been connected to its Florida counterpart. The orange juice industry began operations in Brazil in 1962, when a severe freeze in Florida caused a shortage in the US market. At the time there was no significant international market for FCOJ and Brazilian production was thus targeted to the US market. US companies arrived in Brazil with capital and technology, and formed strategic alliances with packinghouse owners and traders that had access to orange growers. The first exports of Brazilian juice totaled 5,300 tons in 1963.

Florida dominated world orange juice production until the 1980s when a series of freezes during that decade provided a major boost to São Paulo producers.[10] The area planted with oranges in Florida decreased from about 250,000 ha in the late 1970s to 150,000 ha in the late 1980s.

In the 1970s and 1980s, the São Paulo industry enjoyed considerable expansion. In addition to ideal growing conditions for oranges, production costs were relatively lower due to low land and labor costs. The area planted with oranges increased from 118,000 ha in 1961 to 575,000 in 1980 and reached 900,000 in 1990. Fostered by the FCOJ export boom, the number of orange growers and processors increased. Some of these processors were owned by multinational corporations, but most were growers that decided to vertically integrate downstream into processing. These domestic juice processors included cooperatives and family-owned firms. FCOJ exports soared to 700,000 tons in 1980 and surpassed 1 million tons in 1991.

But after the crop failures of the 1980s, Florida's orange groves were relocated to the south, in areas less prone to freezing temperatures and production increased again in the 1990s. There was a glut of FCOJ in the market and prices tanked. The crisis led to industry consolidation. The number of producers in São Paulo decreased from 36,000 in 1995 to 21,000 in 2007,[11] while the planted area was reduced from 700,000 to 500,000 ha. There was also substantial consolidation among processors. The main industry consolidators were two family-owned Brazilian producers – Citrosuco and Cutrale – who invested in processing plants in the 1960s and grew organically and by acquisitions in the 1970s and 1980s. Cargill and Louis Dreyfus Commodities (LDC) entered the industry in the 1980s by acquiring existing plants from domestic

Table 4.5 Growth of the Brazilian orange juice industry (1961–2010)

Orange production[a]	1961	1970	1980	1990	2000	2010
World	15,976,472	24,930,824	40,014,509	49,705,740	63,833,109	69,461,798
Brazil	1,761,768	3,099,440	10,891,814	17,520,520	21,330,258	18,503,139
Brazil share	11%	12%	27%	35%	33%	27%
United States	4,583,450	7,278,428	10,733,810	7,026,000	11,790,680	7,477,924
US share	29%	29%	27%	14%	18%	11%

Orange juice[b]	1985–1986	1990–1991	2000–2001	2010–2011
World production	1,789,463	1,774,503	2,194,893	2,531,057
Brazil production	603,000	949,000	978,000	1,600,000
Brazil share	34%	53%	45%	63%
US production	486,843	623,267	987,682	660,075
US share	27%	35%	45%	26%
World exports	1,257,528	1,344,911	1,374,044	1,517,505
Brazil exports	699,000	989,000	1,075,000	1,185,000
Brazil share	56%	74%	78%	78%
US exports	50,191	69,012	87,193	150,797
US share	4%	5%	6%	10%

[a]Orange production in tons obtained from FAOSTAT.
[b]Frozen concentrated orange juice (FCOJ) production and exports in tons from the USDA.
Source: Elaborated by the author based on data from FAO (2015) and USDA (2015).

Table 4.6 Industry structure in the São Paulo orange juice industry

	Number of processing plants	Number of crushing machines	Share of crushing capacity (%)
Citrosuco[a]	3	312	29.3%
Cutrale	5	290	27.3%
LDC	3	214	20.1%
Citrovita[a]	3	188	17.7%
Other firms[b]	6	60	5.6%
Total	**20**	**1,064**	**100%**

[a]Citrosuco and Citrovita announced their merger in 2010.
[b]Six single-plant processors.
Source: Elaborated by the author based on data available in Neves (2010).

processors and cooperatives. In 2004 Cargill exited the industry and sold its FCOJ assets to industry leaders Cutrale and Citrosuco. In 2010 two of the top-four processors (Citrosuco and Citrovita) announced their merger. The current industry structure is an oligopoly with the three largest firms controlling 95% of the processing capacity. Two of these firms are domestic firms (Citrosuco and Cutrale) and the third-largest processor is LDC, a multinational trading company. Six small, single-plant firms have 5% of the industry processing capacity (Table 4.6).

4.3.1 Explaining the Competitiveness of the São Paulo Orange Juice Industry

While the Florida crop failures of the 1980s provided the "big push" for the development of the São Paulo orange juice industry, they cannot explain how the state achieved international competitiveness over time. Two interrelated factors that help to explain the competitive advantage of the São Paulo industry include growing conditions and productivity gains. As mentioned above, São Paulo offers ideal weather and soil conditions for orange production, with relatively lower production costs and low supply variability. Figure 4.4 shows very clearly that productivity in Brazil is much more stable compared to the US due to more predictable weather patterns. In addition to the several freezes of the 1970s and 1980s, Florida's productivity has suffered the impact of four hurricanes in 2004 and 2005. According to FAO data, land productivity of oranges in Brazil was relatively flat at around 380 boxes[12] per hectare, a level similar to world productivity but significantly below the US productivity levels between 1961 and 1979 (Figure 4.4). Since the 1980s, orange

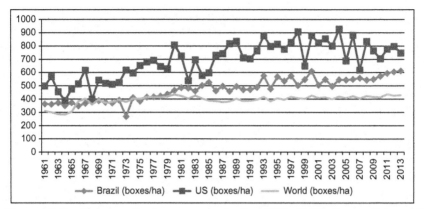

Figure 4.4 Productivity gains in orange production (1961–2013). *Source: Elaborated by the author based on data from FAO (2015).*

productivity in Brazil has increased from 400 to 600 boxes per hectare, catching up with the US average of 750 boxes.

But the FAO data do not tell the whole story, because they include oranges produced for the fresh market and other production regions outside São Paulo and Florida. A different data set shows that the land productivity of orange groves in São Paulo grew at an average annual rate of 2.2% between 1995 and 2013, while it decreased by 1% per year in Florida. As a result, land productivity of oranges in São Paulo averaged 850 boxes per hectare, compared to 800 boxes per hectare in Florida, for the 2011–2012 and 2012–2013 crop years (Figure 4.5). The hurricanes that hit Florida in 2004 and 2005 not only caused damage to orange groves but also helped spread orange tree diseases that negatively affected productivity. The planted area in Florida was reduced from 240,000 ha in 2003 to 175,000 ha in 2013. After reaching an average of more than 1,000 boxes per hectare in 2004, productivity decreased to 770 boxes in 2012–13.

Increased productivity of orange groves in São Paulo is related to the adoption of modern agronomic practices, such as irrigation and densification (i.e., planting more trees per area), stricter rules for the production of seedlings that are free of diseases, better orchard management techniques, and migration of production to areas less prone to diseases.[13] Perhaps more importantly, there has been significant exit of less efficient producers as the FCOJ industry is highly volatile and crisis prone. The crises of 1994–1995, 1999–2000, and 2008–2009 led to significant grower exit. The fact that there are alternative and profitable land uses in

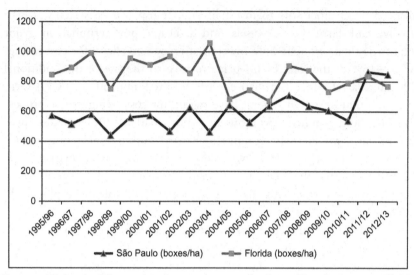

Figure 4.5 Productivity gains in orange production: São Paulo vs. Florida (1995−2013). *Source: Elaborated by the author based on data available in Neves (2010) and the Florida Department of Citrus (2014).*

the state − including sugarcane and eucalyptus − suggests that only farmers with high levels of productivity will maintain or grow orange groves.

In addition to productivity gains at the farm level, industry organization has also played a significant role in the increased international competitiveness of the Brazilian orange juice industry. The first and arguably the most important industry-wide initiative was the creation of Fundecitrus in 1977. Fundecitrus is an industry association funded by growers and processors to control the spread of plant diseases and disseminate best orchard management practices among growers. Fundecitrus focuses on applied research to generate information and technologies to detect, control, and prevent the incidence of diseases in cooperation with public agencies. The fund also invests in genetic improvement to develop plants more resistant to pests and diseases.

Another fact that contributed to industry competitiveness is the large size and economies of scale of processors. As mentioned before, three processors control 95% of industry capacity and have access to state-of-the-art orange-crushing technology. Because the industry has been export-driven since the 1960s, processors have developed the capability to adapt juice blends and quality to specific customer needs in different markets. In addition to technical efficiency in plant operation, size allowed these processors to invest in bulk transportation systems, which involve tank-farm trucks, vessels, and dedicated port terminals in major export destinations. The bulk transportation system allows for cost savings of 10–15% of the final FCOJ price relative to the use of the traditional 200-liter barrel. Since Brazilian exports are predominantly in FCOJ form and bulk transportation systems have significant cost savings, these processors also hold dominant positions in export markets. Other orange processors have two alternatives: rent larger firms' bulk transportation systems or explore the small but growing domestic orange juice market. The Brazilian industry, therefore, has its main competitive advantage in logistics as competitors do not have sufficiently large scale to exploit bulk transportation systems.

The three leading companies in Brazil are also key players in the Florida industry. In the 1990s the four leading firms in the Brazilian orange juice industry at the time – Cutrale, Citrosuco, Cargill, and Dreyfus – started operations in Florida by acquiring existing plants formerly operated by US companies. The explicit motivation for this strategic movement was the increasing difficulties that these firms faced in accessing the US market, the world's largest in terms of orange juice volume. In addition to the trade barriers in the US market, Brazilian companies developed distinctive efficiencies in orange processing that justified these acquisitions. This capability could not be fully explored with plants located in São Paulo as trade barriers and increased consumption of not-from-concentrate (NFC) juices protected Florida production. With the entry of São Paulo-based processors in Florida, the two largest orange juice processors (Minute Maid and Tropicana) exited their orange juice crushing operations to focus on marketing consumer products. After the acquisition of the Coca-Cola plants in Florida, Cutrale became a major supplier to the Minute Maid brand, while Citrosuco signed a supply agreement with Tropicana, the orange juice division of PepsiCo. Brazilian processors are now able to deliver FCOJ and NCF juice year-round to clients across the world, including in Florida. Counting on a reliable and efficient orange juice supply from both Florida and São Paulo, the US beverage companies can focus on their core business in order to fully explore competencies in marketing, distribution, and branding.[14]

In addition to economies of scale, downstream vertical integration, and internationalization, the São Paulo–based processors are also involved in seedling and orange production. Both Cutrale and Citrosuco produce in-house 100% of the seedlings used to renovate existing groves or to plant oranges in new areas. Similarly to sugarcane, oranges are non-storable and highly perishable and must be processed as soon as possible after harvest. Orange groves must be located near processing plants to minimize transportation costs and fruit spoilage. The volume produced by growers and crushed by processors to produce juice is much greater than alternative sources of supply and demand for oranges, making assets dedicated to each other. Processors, therefore, require greater control over production decisions at the farm level – a situation not unlike the sugarcane case.

Processors produce in their own orchards 30–35% of the oranges used in processing, and less than 10% of the processed oranges are acquired in the spot market. The bulk of the processed oranges are acquired from independent growers by means of several types of production and marketing contracts. These contracts are based on boxes delivered (quantity) and do not consider quality attributes that could foster better coordination between processors and growers. Despite the increased use of contracts with pricing formulas based on international FCOJ prices, the majority of the marketing contracts are still based on a fixed price, which increases bargaining costs between processors and growers. In contrast to the sugarcane case, where an organization representing processors and growers sets a base price for sugarcane every year based on technical and market considerations (Consecana), the orange juice industry in São Paulo has been marked by a history of friction between producers and processors.[15] Contracting and pricing issues aside, vertical coordination between growers and processors ensures timely delivery of oranges to crushing plants to maximize operational efficiency and juice quality.

4.4 SUMMARY

This chapter continued our analysis of alternative forms of organization in the Brazilian agrifood sector with a focus on the role of vertically integrated agribusiness. We described the organization of two export-oriented sectors – sugarcane and orange juice – that are clustered in the state of São Paulo in the southeastern region of the country. The dominant form of organization in these two sectors is large, vertically integrated processors that are involved in (almost) all stages of the value chain linking farms to end consumers. In addition to vertical integration upstream and downstream in the value chain, these processors also have contractual agreements with independent sugarcane and orange growers and several forms of collaboration with other organizations in logistics, distribution, and marketing. The sugarcane sector in particular also benefits from industry-wide organizations such as CTC, UNICA, and the Consecana arrangement. These organizations provide public goods to industry participants, including research and development, certification initiatives, standard formula pricing for sugarcane, and engagement with outside stakeholders that contribute to productivity gains, more sustainable industry practices, and lower transaction costs between industry participants. This chapter showed in detail how the organization of these two value chains is a major factor explaining how Brazil achieved high levels of productivity and became the world's largest exporter of sugar, ethanol, and orange juice.

This chapter also described the evolution of the two sectors since the 1970s. The sugarcane industry received a big push from the *Pro-alcohol* policies between 1975 and 1985 but the sector has operated in a liberalized environment since industry deregulation in the 1990s. The orange juice industry, in turn, received a big push from crop failures in Florida in the 1980s but has always operated without any favorable public policy treatment. This chapter analyzed major structural changes occurring in both sectors since the 1990s, including consolidation, internationalization, and ownership structure changes. As a result of these changes, processors consolidated supply, increased scale, adopted best practices in management and governance, and changed their organizations and business models to gain competitive advantage. These structural changes adopted by industry participants also help explain how the sugarcane and orange juice industries adapted to changing market conditions and continued to gain international competitiveness.

NOTES

1. Sections 4.2.1 and 4.2.2 of this chapter build on and update information from Chaddad (2010).
2. TRS is a function of sugarcane production per hectare and the volume of TRS per kg of sugarcane. It is the most important measure of efficiency in a sugarcane mill as it represents 70% of the total production cost of sugar and ethanol.
3. Refer to Martines-Filho et al. (2006) and Moraes and Zilberman (2014) for detailed analyses of the effects of public policy on the evolution of the sugarcane industry in Brazil.

4. Since the operating performance of a processor is mostly affected by agricultural practices and technologies that determine the productivity and quality of sugarcane, which represents about 70% of the production costs of sugar and ethanol, the sustained capacity to improve sugarcane productivity is the most important factor determining business survival and growth in the industry. A recent empirical study shows the operating performance improvements of sugarcane processors following the acquisition wave of the post *Pro-alcohol* period (Mingo and Khanna, 2013).

5. Moura and Chaddad (2012) provide a description of Bonsucro, a multistakeholder initiative that develops sustainability standards for sugarcane production and certifies processors that operate in compliance with these standards. By February 2015, Bonsucro had certified 46 sugarcane mills — 41 in Brazil, four in Australia, and one in Honduras (www.bonsucro.com).

6. Using the jargon of Nobel laureate Oliver Williamson, the transaction between sugarcane growers and processors is characterized by high degrees of locational and temporal asset specificities, dedicated assets, high frequency, and uncertainty, which in turn lead to bilateral dependency and the risk of hold-ups and other opportunistic behavior. Hence, the transaction tends to be governed by hierarchy (Williamson, 1985).

7. This hybrid ownership model was first introduced by dairy cooperatives in Ireland in the 1980s and became known as the Irish model (Chaddad and Cook, 2004). A similar model was adopted by Itambé, the largest dairy cooperative in Brazil.

8. This volume includes sugarcane processed by Copersucar members (about 95 million tons) and product sourced from non-member processors (about 40 million tons).

9. Raizen does not own land; rather it leases and operates farms that are owned by Radar, a subsidiary of Cosan S.A.

10. According to the Florida Department of Citrus, significant freezes in the Florida citrus belt were recorded in 1957, 1962, 1971, 1977, 1981, 1982, 1983, 1985, and 1989.

11. These data refer to numbers of agricultural production units as defined by the office of the secretary of agriculture in São Paulo. Note, however, that one single producer can operate more than one production unit. Neves (2010) estimates that there are 12,500 growers in São Paulo with the largest 10% of them owning 80% of the orange trees in the state.

12. One box of oranges has a net weight of 90 pounds (40.8 kg).

13. Average tree density in orange groves doubled from 250 trees per hectare in 1980 to 500 trees/ha in 2000, while new orange groves are planted with up to 850 trees/ha. More than 130,000 hectares are irrigated in São Paulo, about 30% of the total planted area with orange trees (Neves, 2010).

14. A detailed analysis of the entry of Brazilian orange juice processors in Florida is provided in Azevedo and Chaddad (2006).

15. Starting in 2009, representatives of growers and processors started a process of developing a pricing formula for oranges based on the Consecana model. The first meeting of the Consecitrus organization occurred in October 2014.

REFERENCES

Azevedo, P.F., Chaddad, F.R., 2006. Redesigning the food chain: trade, investment and strategic alliances in the orange juice industry. Int. Food Agribus. Manage. Rev. 9 (1), 18–32.

Chaddad, F.R., 2010. UNICA: challenges to deliver sustainability in the Brazilian sugarcane industry. Int. Food Agribus. Manage. Rev. 13 (4), 173–192.

Chaddad, F.R., Cook, M.L., 2004. Understanding new cooperative models: an ownership-control rights typology. Rev. Agric. Econ. 26 (3), 348–360.

Florida Department of Citrus, 2014. Citrus Reference Book. Available from: <www. fdocgrower.com> (accessed 02.02.15.).

Food and Agriculture Organization of the United Nations – FAO, 2015. FAOSTAT Database. Available from: <www.faostat.fao.org> (downloaded 22.01.15.).

Instituto Brasileiro de Geografia e Estatística – IBGE, 2006. Censo Agropecuário, Brasilia, DF. Available from: <www.ibge.gov.br> (downloaded 09.11.14.).

Martines-Filho, J., Burnquist, H., Vian, C.E.F., 2006. Bioenergy and the rise of sugarcane-based ethanol in Brazil. Choices 21 (2), 91–96.

Ministério da Agricultura, Pecuária e Abastecimento, 2013. Anuário Estatístico da Agroenergia, Brasília, DF. Available from: <www.agricultura.gov.br> (accessed 02.02.15.).

Mingo, S., Khanna, T., 2013. Industrial policy and the creation of new industries: evidence from Brazil's bioethanol industry. Ind. Corp. Change 23 (5), 1229–1260.

Moraes, M.A.F.D., Zilberman, D., 2014. Production of Ethanol from Sugarcane in Brazil: From State Intervention to a Free Market. Springer.

Moura, P.T., Chaddad, F.R., 2012. Collective action and the governance of multistakeholder initiatives: a case study of Bonsucro. J. Chain Netw. Sci. 12 (1), 13–24.

Neves, M.F., 2010. O Retrato da Citricultura Brasileira. Markestrat, Ribeirão Preto, SP.

Neves, M.F., Trombim, V.G., 2014. A Dimensão do Setor Sucroenergético: Mapeamento e Quantificação da Safra 2013–14. Markestrat, Ribeirão Preto, SP.

US Department of Agriculture, 2015. Production, Supply and Distribution Database. Available from: <apps.fas.usda.gov/psdonline> (accessed 02.02.15.).

Williamson, O.E., 1985. The Economic Institutions of Capitalism. Free Press, New York.

CHAPTER 5

Agriculture in the Cerrado: Large-Scale Farming and New-Generation Cooperatives

Contents

5.1 INTRODUCTION

We saw in Chapter 1 how Eugênio Pinesso, a peasant farmer from Paraná, was able to succeed as a commercial farmer in the southern region. Despite his success, he decided to try his luck in the agricultural frontier and acquired his first farm in the cerrado in the state of Mato Grosso do Sul (MS) in 1976. In 1983 Eugênio sold all his land in Paraná to acquire more land in Mato Grosso and the family moved to Campo Grande, MS. The Pinesso Group has expanded since then in Mato Grosso and other states in the Brazilian cerrado and planted 117,000 ha in 2014. The story of the Pinesso family is not unique as the cerrado was conquered by farmers, ranchers, and entrepreneurs who migrated primarily from southern and southeastern Brazil.[1] In this chapter we explain how agriculture has developed in the Brazilian cerrado since the 1970s.

F. Chaddad: The Economics and Organization of Brazilian Agriculture.
DOI: http://dx.doi.org/10.1016/B978-0-12-801695-4.00005-7

As described in Chapter 2, the Brazilian cerrado covers 200 million hectares and cuts across many states. Until the 1970s, the cerrado was relatively "empty" and considered to be of limited value for agricultural production. With the development of new agricultural technologies adapted to the poor soils of the cerrado and crop varieties able to grow in low latitudes, the cerrado became a breadbasket. Since the cerrado is such a vast biome, this chapter focuses on how agriculture developed in the state of Mato Grosso, which has become the leading crop producer in the country in the last decade.

5.2 EVOLUTION OF FARMING IN MATO GROSSO

Table 5.1 shows the evolution of farming in Mato Grosso since the 1970s. In 1975 total land in farms was about 22 million hectares or 24% of the state's landmass. But only 11.7 million hectares or 54% of the land in farms were actually cleared of natural forests and used in production, of which 8.6 million hectares were in natural pastures, 2.6 million hectares in planted pastures, and 500,000 ha in temporary and permanent crops. At that time

Table 5.1 Evolution of agriculture in Mato Grosso, Brazil (1975–2006)

	1975	1985	1995–1996	2006
Number of farm establishments	56,118	77,921	78,762	112,987
Total land in farms (ha)	21,949,146	37,835,651	49,839,631	48,688,711
Share of state landmass in farms (%)	24%	42%	55%	54%
Average farm size (ha)	391	486	633	431
Area cleared in establishments (ha)	11,767,758	18,559,984	24,471,635	28,559,105
Permanent crops (ha)	42,174	136,605	169,734	408,550
Temporary crops (ha)	459,093	1,992,838	2,782,011	6,018,182
Natural pastures (ha)	8,640,861	9,685,306	6,189,573	4,404,283
Planted pastures (ha)	2,602,607	6,719,064	15,262,488	17,658,375
Natural forests in establishments (ha)	7,101,035	14,126,813	21,475,765	19,106,923
Share of land in farms cleared (%)	54%	49%	49%	59%
Number of employed farm workers	263,179	359,221	326,767	358,336
Number of tractors	2,643	19,534	32,752	42,330
Number of beef cattle	3,110,119	6,545,956	14,438,135	20,666,147

Source: IBGE (2006).

the state did not have adequate infrastructure and was isolated from the rest of the country. In addition, the agricultural technologies adapted to the tropics developed by EMBRAPA and other research institutes were yet to be developed and disseminated. Thus farming activities in the state of Mato Grosso were predominantly extensive cattle ranching and staple crops for subsistence, both with low levels of productivity.

This situation started to change with a set of policies implemented by the military regime that aimed to integrate the Legal Amazon[2] to the rest of the country with the enactment of the National Integration Plan (PIN) in 1970. These policies included infrastructure development (primarily roads), tax exemptions for enterprises investing in the region, and subsidized credit. The highway system grew from 400 km in 1968 to 56,000 km in 2001, of which 20,000 km were paved.[3] The early beneficiaries of these policies were cattle development projects. By 1980, the federal planning agency for the region (SUDAM) had approved 469 cattle-related projects involving US$565 million.[4] In addition to these large-scale cattle ranching projects, PIN also included development initiatives as a response to a widespread drought in the northeast region, which increased the incidence of poverty and famine among peasant farmers. State-led colonization projects were designed to settle poor farmers and landless individuals in the Legal Amazon. Major state-led colonization projects settling more than 5 million hectares were implemented in the states of Amazonas, Acre, Mato Grosso, Pará, and Rondônia in the 1970s and 1980s.

Private colonization projects organized by colonization firms and cooperatives were also relevant agents of agricultural development in the Legal Amazon. Not unlike the cases of the Dutch and German colonization projects of the 1950s in Paraná described in Chapter 3, private colonization entities surveyed, demarcated, and occupied land, built infrastructure, opened roads, developed urban areas, and provided basic health and education services to smallholders that came primarily from the southern region. Thirty-five private enterprises organized 104 settlement projects and colonized almost 4 million hectares of land in Mato Grosso between 1970 and 1990. Private colonization projects in Mato Grosso represented 39% of the total area colonized in the Legal Amazon.[5]

As a result of these policies to develop the Legal Amazon, which attracted farmers, ranchers, and land developers from all over the country, the number of farm establishments in Mato Grosso increased from 56,000 to 78,000 and the total land in farms almost doubled from 22 to 38 million hectares between 1975 and 1985 (Table 5.1). During this period, the area actually used in production increased from 11.7 to 18.5 million hectares, with notable increases in planted pastures (from 2.6 to 6.7 million hectares) and temporary crops (from 460,000 to 2 million hectares). Many critics of these policies argue that this first phase of agricultural development in the cerrado led to excessive land clearing, environmental degradation, and increased land concentration (see, for example, Klink and Machado, 2005, Mueller, 2003, and Walker et al., 2009). The poor soils of the cerrado could not sustain adequate levels of productivity after 3 years of being used in production. Without investments in soil fertility, natural pastures would degrade and row crops would not produce enough to cover costs.

The integration and development policies initiated by the military government were, however, short-lived. In 1979 Brazil experienced the effects of the second oil crisis, which severely affected the ability of the government to invest in proactive development policies. We saw in Chapter 2 how government expenditure in agricultural policy reached its peak in the 1980s and was significantly reduced after that. Following restoration of democracy in 1985, subsequent policies related to the development of the Amazon region cut subsidies and tax incentives to agricultural projects and started to pay more attention to environmental issues, the protection of indigenous land rights, and land reform. Substantial tracts of cerrado and tropical rainforests in the Legal Amazon are now off limits to agricultural expansion and remain protected in nature conservation units and indigenous reserves. Land reform settlements also started to be implemented in the 1990s to address the issue of land concentration.

A second major push to agricultural development in the cerrado occurred after the macroeconomic reforms of the 1990s. We explained in Chapter 2 how the Real Plan of 1994 played a critical role in agricultural development throughout the country with currency stabilization and the control of inflation. Initially, the Real Plan led to a crisis in agriculture as a result of an overvalued currency but the currency devaluation of 1999 under a free-floating exchange rate policy — coupled with increased international demand for agricultural commodities in the 2000s — provided a massive boost to agricultural development. Also, since the mid-1980s, new agricultural technologies started to become available to farmers in the cerrado, which led to the development of commercial agriculture and ranching with increasing levels of productivity.

Between 1985 and 2006, the area with planted pastures in Mato Grosso increased from 6.7 to 17.6 million hectares and the number of beef cattle increased from 6.5 to 20.6 million head. The beef cattle herd reached 28 million head in 2014 with beef production representing 20% of the state's gross value of agricultural production. The improvement of forages coupled with appropriate soil management practices solved the issue of pasture degradation and significantly improved the economic prospects of cattle ranching in the cerrado. The livestock industry also benefited from government actions to control foot and mouth disease and to eliminate sanitary barriers to Brazilian beef in foreign markets. With the expansion and increased productivity of cattle herds in the cerrado, Brazil has become the second-largest producer and the leading exporter of beef in the world.

Even more impressive than the growth of the livestock sector has been crop production expansion – in particular, soybeans, corn, and cotton. Between 1985 and 2006, the area planted with temporary crops in Mato Grosso tripled from 2 to 6 million hectares (Table 5.1). Recent data show that crops continued to expand in the state, with the planted area reaching 13 million hectares and total crop production of almost 48 million tons in 2014 (Table 5.2). Agricultural production growth resulted from planted area expansion and productivity gains, which increased from 1.4 tons per hectare in 1977 to 3.6 tons in 2014. Mato Grosso surpassed Paraná as the country's largest producer and currently accounts for 25% of the domestic grain and oilseed production. The state produces 30% of the Brazilian

Table 5.2 Evolution of crop production in Brazil and Mato Grosso (1977–2014)[a]

	1976/ 1977	1984/ 1985	1994/ 1995	2004/ 2005	2013/ 2014
Brazil					
Planted area (1,000 ha)	37,314	39,693	38,539	49,068	56,988
Crop production (1,000 tons)	46,943	58,143	81,065	114,695	193,386
Productivity (kg/ha)	1,258	1,465	2,103	2,339	3,393
Mato Grosso					
Planted area (1,000 ha)	2,238	1,561	3,278	8,564	13,323
Crop production (1,000 tons)	3,046	2,642	7,617	24,731	47,703
Productivity (kg/ha)	1,361	1,693	2,324	2,878	3,580

[a]Includes 15 crops.
Source: CONAB (2015).

soybean crop, 23% of the corn crop, and 58% of the cotton crop. In 2014 crop production represented 74% of the state's gross value of agricultural production, with soybeans alone accounting for 50% of the total value.

5.3 PRODUCTIVITY GAINS

As we saw in the previous section, crop production in Mato Grosso experienced tremendous growth in the last four decades, increasing from 3 million tons in the mid-1970s to almost 48 million tons in 2014 (Table 5.2). This production growth of about 10% per year since 1976 was due to planted area expansion (at a rate of 6.8% per year) and land productivity gains (3.1% per year). Soybean production growth reached an annual growth rate of 15% with land productivity gains of 2% per year in the same period. Figure 5.1 shows that soybean productivity in Mato Grosso surpassed the national average in 1980 and has remained above it since then. The productivity of Mato Grosso producers has also rivaled the yield levels achieved by their counterparts in Paraná and the United States. In this section we will take a closer look at how Mato Grosso achieved such impressive production growth and productivity gains with a focus on soybeans.

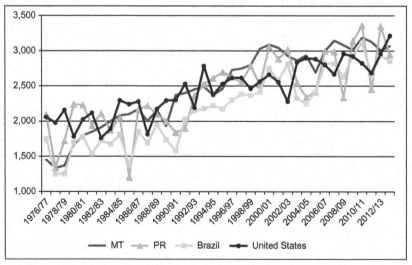

Figure 5.1 Evolution of soybean yields in Brazil and the United States in kg/ha (1977–2014). *Source: CONAB (2015) and NASS (2015).*

Soybeans were first introduced in Brazil in 1882, but the crop started to gain economic relevance only in the 1970s. We saw in Chapter 2 that commercial production of soybeans started in the southern region with the introduction of cultivars bred from germplasm imported from the southern United States and the improvement of soil fertility with lime and fertilizer application. In the 1970s, Brazilian researchers in public and private research institutes started to breed new cultivars better adapted to the local growing conditions. We described in Chapter 3 the role of cooperatives and public—private research partnerships in the development and dissemination of new technologies to family farmers in the southern region. Until the 1980s, 90% of the national soybean crop was produced to the south of the Tropic of Capricorn (23°S latitude) primarily in the southern region.

In the early 1980s, researchers coordinated by the EMBRAPA soybean center in Londrina, PR, started to breed new cultivars that would allow soybean to be produced in lower latitudes. The main challenge that they needed to overcome was to understand how the soybean plant would react to the shorter day conditions of the tropical region. The strategy followed was to identify genes in the available cultivars planted in southern Brazil that expressed late flowering under short day conditions, which is known as the long juvenile trait. New cultivars were bred with a range of sensitivity to day length that adapted well to diverse growing conditions in lower-latitude regions. Before being introduced on a commercial scale, these new cultivars were tested in private and public experimental fields across the country. In Mato Grosso, the leading organization that worked with EMBRAPA soybean breeders was the Mato Grosso Agricultural Research Foundation (Fundação Mato Grosso — FMT), which we describe below. With the introduction of these new cultivars, and appropriate soil fertility and management techniques, soybean production in Mato Grosso increased from 230,000 tons in 1980 to 3.7 million tons in 1989.

With the growth of soybean production in the cerrado, new disease and pest problems emerged in the 1990s that negatively affected soybean yield, including nematodes, stem canker, and frog-eye leaf spot. Again, Brazilian researchers had to rise to the task and breed new cultivars resistant to plant diseases. The introduction of new cultivars in the 1990s also diversified the genetic pool of commercial seeds, which provided another boost to soybean yields. As a result of these agricultural R&D efforts, soybean yield in Mato Grosso increased from 1.7 tons in 1980 to 3 tons per hectare in 2000. By the early 2000s, however, a new disease, Asian soybean rust, was introduced into the country. Asian rust affected soybean yields, which declined to about 2.5 tons per hectare between 2003 and 2006, but also increased production costs as disease control required multiple applications of fungicides.[6]

As a response to the threat posed by Asian rust, a new cropping system emerged along with the introduction of new cultivars. This new cropping system was based on early planting of soybeans that had a better chance of escaping the disease and required less use of fungicides. New cultivars with the indeterminate growth trait and a shorter reproductive period allowed earlier planting of soybeans. These new cultivars sacrificed yield potential of soybeans but allowed better control of Asian rust and double cropping.[7] The short-cycle, indeterminate soybean cultivars introduced in the mid-2000s also required the use of minimum tillage systems to improve soil fertility and the adoption of the herbicide resistance trait developed by Monsanto to control weeds. With this new "technological package," soybean-planted area and production continued to expand in Mato Grosso in the late 2000s. Soybean yields recovered the ground lost to the spread of Asian rust and returned to 3 tons per hectare after 2007.

Despite lower yield potential, early planting of soybeans with short-cycle cultivars has become very popular among producers in Mato Grosso as it allows a second crop in the same field — such as corn, cotton, pasture, or some cover crop — after soybean is harvested. The area planted with corn as a second crop in Mato Grosso increased tenfold from 300,000 ha in 2000 to 3.3 million hectares (about 40% of the soybean area) in 2014, while corn yield increased from 2 tons to 5.5 tons per hectare with the introduction of new corn hybrids adapted to this new production system. In 2014 the state produced 18 million tons of corn, of which 17.6 million tons were produced as a second crop following soybean harvest. Another popular crop rotation after the soybean harvest is with planted pastures, which allows the fattening of livestock — a system known as crop–livestock integration. As explained in Chapter 2, adequate crop rotation of soybeans with corn, pastures, or cover crops is the best agronomic practice to control the incidence of pests and diseases in the tropics.

5.3.1 The Mato Grosso Agricultural Research Foundation (FMT)

It was 1990 when a new researcher called Dario Hiromoto arrived at the EMBRAPA soybean center in Rondonópolis, MT, to test new seed varieties and develop soybean production systems adapted to the diverse growing conditions in the state.[8] At that time, EMBRAPA had an agreement with a state research institute called EMPAER to test its new soybean varieties in the state but the available resources were meager. Dario convinced producers in different regions across the state to allow him to set up experimental fields in small plots in their farms. One of these producers in Campo Novo dos Parecis, called Sérgio Stefanello, explained Dario's impact as follows: "Before the arrival of Dario in Mato Grosso, it was the researcher on one side and the producer on the other. There was no communication. Research solutions developed by federal or state institutes took 3 to 4 years to become available to farmers. With Dario, the communication improved tremendously and research results arrived in the fields in real time" (Fundação de Apoio à Pesquisa Agropecuária de Mato Grosso, 2010, p. 37).

Dario did not last long at EMBRAPA. Agriculture was developing at a fast pace in Mato Grosso and producers were facing increasing challenges related to soil fertility, plant nutrition, and the emergence of new plant diseases and pests, including nematodes and stem canker. In addition, the soybean cultivars developed in other states were ill-adapted to the local growing conditions. With ever-decreasing investments in public research institutes like EMPAER in the early 1990s, Dario and his team did not have enough resources to meet these challenges. His solution was to establish a research foundation called *Fundação de Apoio à Pesquisa Agropecuária de Mato Grosso* (FMT) to be funded by the producers. His idea received the support of two influential farm leaders, Gilberto Goellner and Blairo Maggi, who organized a group of 23 producers to initially fund FMT. These producers were concerned about the future of soybean production in the state and realized the need for a research program that could deliver timely solutions to the problems of soybean production and generate higher-yield varieties adapted to the cerrado conditions. FMT was established in December 1993 and Dario was named its first executive director, overseen by a board of directors comprised of producers.

In its first years of operation, FMT focused on developing and adapting new soybean cultivars to the diverse growing conditions of the state. FMT signed an agreement with EMBRAPA to use its germplasm and test new soybean seeds in experimental fields across the state. In addition to improving yields, the soybean breeding program of FMT aimed at introducing seed varietals resistant to diseases. In 1996 FMT brought to market nine soybean cultivars resistant to the cyst nematode and stem canker. According to Blairo Maggi, "Dario was able to find a research solution in 3—4 years for a problem that would normally require 10—12 years of research." By 1998, 77% of all soybean seeds produced in the state were cultivars developed by the EMBRAPA—FMT partnership (Nassar, 1998).

In 1997 the agreement with EMBRAPA was expanded to cotton. In addition to the soybean and cotton breeding programs in collaboration with EMBRAPA, FMT also conducted research in plant nutrition, soil fertility, plant disease control, and development of alternative crops adapted to the cerrado. Dario was as passionate with outreach to producers as he was with research. FMT thus organized several field days and workshops across the state to communicate research findings and solutions to farmers. Some of these field days would gather as many as 4,000 producers.

The research collaboration agreement between FMT and EMBRAPA lasted until 2000. The two organizations had different cultures and dynamics. The issue of how to share the royalties from new seed development also tore the two organizations apart. Following the break-up with EMBRAPA, FMT changed its organizational structure to become more market-oriented and generate enough funding to support its research program. A new corporate entity called Tropical Melhoramento e Genética (TMG) was founded having Unisoja and TGX as the two controlling shareholders with 70% and 30% of the shares, respectively. Unisoja is the entity comprising 58 producers that funded the research foundation (FMT) since 1993. TGX, in turn, is the entity formed by the principal researchers associated with the genetic and breeding programs of TMG, including Dr. Romeu Kiihl, who also left EMBRAPA to join TMG as its research director.[9]

TMG headquarters and research labs are located in Cambé, PR, not far from the EMBRAPA soybean center in Londrina. TMG is responsible for basic research in genetics, biotechnology, and plant breeding. It uses the germplasm developed over the last 20 years by FMT to identify, evaluate, and select promising new cultivars to be tested in 50 FMT experimental fields across Mato Grosso and other states (Mato Grosso do Sul, Goiás, Minas Gerais, and Bahia). FMT continues to be responsible for new cultivar testing, applied research in production systems, and communicating research results to producers. In other words, the foundation focuses on providing public goods to farmers in the cerrado.

In this new corporate structure, TMG is able to charge royalties for the seeds multiplied and commercialized by a network of 60 seed firms and thus generate funding for its research program. About 95% of the royalties generated by the TMG seed business are reinvested in the research program. Despite increasing competition from multinational seed companies since the early 2000s, when Brazil approved the research, development, and distribution of genetically modified (GM) seeds, TMG continued to innovate and maintain relevant market shares in its trade territory. In 2005 and 2007, TMG introduced new soybean cultivars resistant to nematodes and in 2008 it introduced a super-precocious cultivar that is resistant to rain during harvest. Also, in 2008, TMG was the first seed company to develop soybean varieties that are resistant to Asian rust and require less use of fungicides. TMG has a strong breeding program in non-GM varieties but it also has licensing agreements with other seed companies — including Monsanto and BASF — to use their biotechnology traits.

According to Francisco Soares Neto, the current chief executive officer (CEO) of TMG, it is very hard for a seed company to survive in Brazil based solely on conventional breeding programs. "Although conventional breeding allows us to develop seeds with the potential to double current soybean yields, it is very hard for the traditional plant breeder to appropriate the value created. The Brazilian law is lax in protecting the plant breeder's rights, but it protects the intellectual property rights of biotechnology firms, as their traits are protected by patent laws. This is why biotech firms have acquired almost all Brazilian seed companies to use their germplasm as vehicle for the commercialization of biotech traits." After the sale of Coodetec to Dow in 2014, TMG is currently the only Brazilian firm with relevance in the seed market, with a 20% national market share and a 45% share in Mato Grosso.

5.3.2 EMBRAPA Agrossilvipastoril

We saw previously how EMBRAPA agricultural research efforts since 1975 and its collaboration with public and private institutes in the 1990s — including FMT in Mato Grosso and Emgopa in Goiás — enabled the development of agriculture in the cerrado. We also described in Chapter 2 how, since the institutional changes of the late 1990s providing enhanced protection of intellectual property rights, EMBRAPA and other public and private seed breeders lost market share to multinational companies. EMBRAPA continues to develop research in plant breeding and genetics, but its impact in the future development of the cerrado might come from other research efforts, primarily in integrated production systems that combine production of crops, livestock, and forests in the same area — known as agrosilvopastoral (ASP) systems.[10]

EMBRAPA's ASP research center was established in 2012 in Sinop, located 500 km to the north of Cuiabá, the state capital. Since then it has recruited an interdisciplinary team of 50 PhDs to conduct applied research to develop ASP systems adapted to tropical conditions.[11] Two large experiments were set up, each utilizing about 100 ha of experimental fields — one based on beef cattle production and the other on milk production, but both integrated with crops and forestry production. The goal of this multidisciplinary research effort is to evaluate the agronomic, economic, and environmental effects of the adoption of several combinations of ASP systems. "Our goal is to show that integrated production systems provide a better solution to the challenges of agriculture in the tropics. Current research shows that integrated production systems have positive effects on soil fertility and control of diseases and pests with better agronomic and sustainability outcomes. We are working hard to estimate the economic benefits of the adoption of ASP systems. If we succeed in developing economically viable ASP systems, it will change the paradigm of food production in the tropics," Dr. João Veloso Silva, the general manager of the research center, told me.

According to Dr. Lineu Domit, who is in charge of the extension and technology transfer activities of the center, there are several potential economic benefits of integrated crop—livestock—forestry systems, including economies of scale in production, more efficient utilization of machinery and labor during the year, income diversification, and reduced risks. In particular, an economically viable integrated production system would allow farmers to decrease their dependence on the soybean—corn rotation. Additionally, he believes that ASP systems based on milk production have the potential to integrate a larger number of smallholders to supply chains. "About 70% of the farmers in Mato Grosso are smallholders with less than 100 ha and peasants in land reform projects. We need to provide them economic and sustainable solutions to increase their income."

5.4 ENTREPRENEURSHIP AND ECONOMIES OF SCALE

In addition to agricultural research conducted by public and private organizations, the development of agriculture in the cerrado would not have occurred without the entrepreneurship of farmers. Based on my field research in Mato Grosso, we explore how farmers were able to make it in the agricultural frontier despite market failures, missing services, poor

Table 5.3 Farm structure in Mato Grosso (2006)

	Number of establishments	Share	Area (ha)	Share	Average area (ha)
0−100 ha	77,786	68.8%	2,641,168	5.4%	34
100−500 ha	21,334	18.9%	4,503,981	9.3%	211
500−2,500 ha	10,052	8.9%	11,316,022	23.2%	1,126
More than 2,500 ha	3,815	3.4%	30,227,539	62.1%	7,923
Total	112,987	100.0%	48,688,710	100.0%	431

Source: IBGE (2006).

infrastructure, and market volatility. This section highlights the role of economies of scale and producers' organizations, such as cooperatives, in overcoming the challenges of farming in the cerrado.

We saw above that Mato Grosso attracted a large number of migrants from several parts of the country — but primarily from the southern region — in the 1970s and 1980s. Unfortunately, a large number of these pioneers did not succeed and were forced to exit agriculture, especially during periods of economic crisis. The most notable crises include the period of hyperinflation of the late 1980s and early 1990s; the crisis of 1994−1995 following the Real Plan and the emergence of new soybean diseases; and the "perfect storm" of 2004−2006 with the arrival of Asian rust coupled with weak commodity prices and exchange rate appreciation. Farmers across the world are used to the volatility of markets and Mother Nature, but periods of crisis have a stronger impact in countries without "safety nets" such as Brazil. In addition, the effects of a crisis are more pronounced in frontier regions such as Mato Grosso because they are distant from the ports and marketing margins tend to be lower. As a result of these several crises, a very large number of farmers had to sell land to pay off debt.

Perhaps not surprisingly, the structure of farming is highly concentrated in Mato Grosso. Data from the last Census of Agriculture show that the average farm size in the state is 431 ha, which is significantly higher than the national average of 64 ha (Table 5.3). About 70% of the producers in the state farm less than 100 ha. These farm establishments include traditional smallholders and peasants who were settled in land reform projects. These are the farmers targeted by the family agriculture program (PRONAF) described in Chapter 2. The total land in smallholder farms adds up to 2.6 million hectares, which is equivalent to 5% of the total agricultural land in the state. Among commercial producers, there are three rough size categories: small (100—500 ha), medium (500—2,500 ha), and large (above 2,500 ha). The 35,000 commercial producers in Mato Grosso farm 46 million hectares or 95% of the agricultural land in the state.

5.4.1 Commercial Producers

Let us take a closer look at some of these commercial producers. Nelson Piccoli came to Mato Grosso from western Santa Catarina, where he used to work as an accountant. He bought a 250-ha farm in Nova Mutum, located 250 km to the north of Cuiabá on the main federal road crossing the state (BR-163), in 1984. He sold this farm in 2001 to acquire a larger farm of 1,630 ha in Sorriso, another 160 km to the north of Nova Mutum on BR-163, which has become the "agribusiness capital" of the cerrado. He paid 40 soybean bags per hectare to acquire the farm from a private colonization firm in installments paid over 5 years.[12] It took his family about 5 years to clear the land and build soil fertility with liming and application of fertilizers, which cost another 40 bags per hectare. He currently plants about 1,000 ha of soybeans every year, double cropping with corn and pastures to grow livestock. Following federal environmental laws, about 480 ha on the farm are set aside as "legal reserve" and "permanent protection area" to preserve natural vegetation and the environment. In contrast to his counterparts in developed countries, he does not receive any subsidies or payments for providing these environmental services.

Mr. Piccoli currently spends most of his time as leader of several producer associations and cooperatives in the state. He serves as financial director of Famato and Aprosoja, two producer associations, and as board director in two agricultural cooperatives and one credit cooperative. The farm is run by his son, Hernandes, who has a college degree in agronomy. According to Mr. Piccoli, "Hernandes makes all major decisions on the farm. He may eventually seek my advice, but he is the ultimate decision maker." I asked him if he plans to expand. "I was fortunate to build a comfortable net worth. I invested 80 bags per hectare to acquire and develop this farm and now it is worth ten times more. I can rent the land to my son for eight soybean bags per hectare,[13] which is more than enough to cover my living expenses. I also realized that growing the farm or buying more farms will not make me happier. I prefer to spend my time leading producer organizations. In the future, our challenge in Mato Grosso is not to plant more hectares, but rather it is to add more value to our agricultural production." We analyze below the role of producer associations and cooperatives in the state.

Another commercial producer whom I visited is Silvésio de Oliveira, who came from Marechal Rondon in western Paraná to Mato Grosso in 1987 with his four brothers. "Our parents farmed 18 hectares in Paraná, growing crops, raising hogs, and milking cows. They were members of C-Vale, a large agricultural cooperative,[14] and were able to earn a decent living in agriculture. But there was no room for my brothers and me." The Oliveira brothers acquired 200 ha in Tapurah, MT, in 1987 from a private colonization firm called Colonizadora Nova Eldorado. They worked hard to develop the land and make a living in agriculture, but had to sell the farm to pay off debt in 1995. "We paid the price of being pioneers in a new region, far away from markets, in very uncertain times. It was a matter of luck. Some producers survived the crisis but many had to leave." The brothers split up and Mr. Oliveira started from ground zero again. He continued to farm on leased land and provided management services to livestock farms in the region with absentee landlords. He benefited from the macroeconomic stability that followed the Real Plan to invest in machinery and on-farm grain storage. "I learned my lesson and only invested when I had enough cash. I did not want to depend on banks anymore. I also started to save 20,000 bags of soybean each year as my insurance policy."

In 2002 one of his landlords sold the beef cattle and formed a crop-share partnership with Mr. Oliveira. They currently adopt an agro-pastoral rotation system, with 1,350 ha of crops in addition to 150 ha of pastures. Mr. Oliveira explained that "it is important to keep the land without soybeans at least for 2 years in a row. During this time we plant cover crops or pastures to raise livestock. Crop rotation is the best strategy to control pests and diseases in our region. We noticed that nematode infestation increases in areas with soybean monoculture, which reduces productivity by 30%." Mr. Oliveira also serves as the local representative of Aprosoja in his municipality. His major role is to organize monthly meetings with farmers to listen to their challenges and participate in monthly meetings in Cuiabá with his counterparts representing farmers from other regions across the state. This information gathered directly from farmers is used by Aprosoja to guide its policy efforts.

Similarly to Messrs. Piccoli and Oliveira, there are about 10,000 commercial producers in Mato Grosso farming an average of 1,100 ha (Table 5.3). They are family farmers[15] who were able to survive several crises since the 1980s and establish themselves as commercial producers. Most of them agree that 500 ha is the minimum efficient scale that a producer needs to farm to be able to acquire the machinery, build the on-farm infrastructure, and have access to modern inputs to make a living in agriculture.

Other producers, however, were able to achieve a much larger scale than this. Table 5.3 shows that there are about 4,000 large commercial producers in Mato Grosso with an average farm size of 8,000 ha. These large producers farm about 30 million hectares or 62% of farmland in the state. According to primary data collected by Agroconsult, a consultancy firm, there are 38 "mega producers" farming more than 30,000 ha in the cerrado region, with several of them farming more than 100,000 ha. One of these producers is the Pinesso Group that developed from the work of Mr. Eugênio Pinesso, described as an example of a successful farmer-entrepreneur in the southern region in Chapter 1.

In 1983, Eugênio took the bold decision to exchange five farms totaling 1,500 ha in Paraná for two farms with 19,600 ha in Campo Verde, MT, located 100 km to the east of Cuiabá. The farms were not yet developed so family members worked hard to clear the land and bring it into production. Eugênio had six daughters and sons and most family members were involved in agriculture. They worked as a team. While some family members worked in developing land into crop production in Mato Grosso, other family members stayed in Mato Grosso do Sul focusing on livestock production, including hogs and beef. One main characteristic of large producers in the cerrado is that they developed as family groups, pooling resources from many family members under the leadership of one trusted individual. With the passing of his wife in 1986, Eugênio started to groom one of his sons, Gilson Pinesso, to take over his role as the family leader. Similarly to his father, Gilson was a hard-working and shrewd businessman, who had earned the trust of his siblings and family members since he started working in the family business in the 1960s. He took over the leadership of the family group in 1988.

In 1993 the Pinesso Group took another bold step with the acquisition of a 61,000-ha farm in Nova Ubiratã, MT, located 150 km to the east of Sorriso in an area without any infrastructure, including roads and electricity. The land was developed over time and required investments in basic infrastructure, including building more than 200 km of roads, bridges, housing, electricity, and a school for rural workers and their families. In 2014 the district developed by the Pinesso Group had a population of 2,000 with a school, health center, pharmacy, supermarket, and service station. The farm produces soybeans, corn, cotton, and sunflower.

In addition, it has facilities to finish 30,000 head of beef cattle in confinement and a hog production system from farrowing to finish for 71,000 hogs. In 2014 the Pinesso Group had ten farms in the cerrado totaling 108,000 ha. It also leased 48,000 ha in the new frontier state of Piauí. Planted area with crops increased from 48,000 ha in 2005 to 117,000 ha in 2014. The area used for livestock production in planted pastures covered 35,000 ha.

In 2012 the Pinesso Group incorporated and changed its name to Produzir S.A., which literally means "to produce." All assets were transferred to the corporate entity and family members received shares in return. Produzir S.A. has six shareholders, each a limited holding company controlled by the six siblings (or their descendants). Gilson explained to me why the family decided to adopt a corporate structure as follows. "We decided to incorporate for two main reasons. The first was to give shares to family members, which would allow exit at fair value. The second was to enable the firm to adopt corporate governance practices and hire professional managers." The board of Produzir S.A. is comprised of six directors nominated by each family holding company (the shareholders) and one independent, professional director. A CEO with professional experience in finance was hired to run the business. Although Gilson was increasingly involved in leadership positions in many producer associations, the main reason why he stepped down from the CEO position was that "we needed someone from outside the family to make business decisions with cold blood and without emotion."

The CEO also plays the role of chief financial officer (CFO) and the company hired two additional senior-level professionals – a chief operating officer (COO) to run the farming operations and a chief commercial officer (CCO) in charge of commercial and marketing decisions. This senior management team is assisted by six mid-level managers – four of them responsible for managing crop operations and two in charge of the livestock operations. The firm has 1,050 employees, with 110 of them involved with some management or administrative function. As a result of these organizational changes, family members – including Gilson – are no longer directly involved with the business. The senior management team receives incentive compensation based on clearly defined goals, including growth and profits.

Another example of a "mega producer" in Mato Grosso is Grupo Bom Futuro (GBF) led by Eraí Maggi Scheffer. Eraí grew up on a family farm in São Miguel do Iguaçu, not far from the Iguazu Falls in western

Paraná. From the age of 6 years he helped his parents on the farm when he was not in school. In the late 1970s he went to Mato Grosso to work on a farm in Juara, 640 km to the northwest of Cuiabá. In 1982 he started to plant soybeans with his two brothers on leased land near Rondonópolis. The family group prospered and in 1993 they were able to acquire their first farm called Bom Futuro. According to Eraí, "when times are good, everything is expensive. The economic crisis of the early 1990s was the perfect opportunity to buy land." In 1994 the family group started to plant cotton, which accelerated growth. Since then they have never looked back.

By the end of 2012 GBF was the largest producer of soybeans and cotton in Brazil. In that year the group planted 230,000 ha of soybeans and 20,000 ha of cotton as first crops, followed by 70,000 ha of corn and 50,000 ha of cotton as second crops. Crop production generated 80% of GBF's revenues, which surpassed BRL 1 billion in 2012. About 10% of total revenues came from livestock operations, which included 90,000 beef cattle and the production of 1,500 tons of fish per year. The family group also had a seed business, which generated 10% of the firm's revenues. GBF is one of the major shareholders of Unisoja, which is the controlling shareholder of FMT and TMG.

Agricultural production is carried out in 40 production units − or clusters of farms ranging from 5,000 to 10,000 ha that share the same infrastructure, machinery, and personnel − located in five regions across Mato Grosso. About half of the land is owned by GBF and the other half is leased from other producers. GBF has 5,000 employees − the majority of whom work on the farms. About 800 families actually live on the farms operated by GBF. The corporate headquarters in Cuiabá houses a team of 200 professional managers.

In a personal interview I asked Eraí how he was able to oversee such a large farming operation. In simple terms, he explained to me that "in agriculture, it is the eye of the owner that fattens the ox." He, together with his two brothers and one brother-in-law, are involved in the business. They constantly visit the farms and develop personal relationships with farm managers and workers. "We have formed our team over the last 30 years. Today we have the sons and daughters of our first employees working for GBF. They have a sense of belonging to the group and we offer a lot of opportunities for personal growth and development."

In addition to a strong organizational culture based on personal relationships, GBF has recently adopted a control system with key performance indicators to monitor the performance of its 40 production units across the state. "We benchmark the performance of each production unit relative to the others, which provides a strong incentive for the teams to perform. The managers of these production units also share information and best practices among themselves." GBF also introduced an incentive compensation system for farm managers based on production and productivity levels achieved by each production unit.

5.4.2 Corporate Farms

Another recent phenomenon that is changing the structure of agricultural production in the Brazilian cerrado is the emergence of corporate farms since the mid-2000s. Corporate farms are increasingly found in Mato Grosso, but especially in the new agricultural frontier of the cerrado known as Mapitoba, a region comprising four states – Maranhão, Piauí, Tocantins, and western Bahia. These new players include publicly traded companies (e.g. SLC Agrícola, Vanguarda Agro, and BrasilAgro), privately held companies controlled by private equity funds (e.g. Agrifirma, Agrinvest, and Tiba Agro), and subsidiaries of multinational trading companies (e.g. Ceagro-Mitsubishi and XinguAgri-Multigrain). These corporate farms with diverse ownership arrangements have three characteristics in common – very large scale, professional management, and access to capital markets. According to industry sources, ten corporate farming entities planted about 1 million hectares in the Mapitoba region in 2013, which represented one third of the total area in production. These corporate entities were attracted to the region because of low land prices relative to other established regions in the cerrado and closer proximity to the ports in northeastern Brazil.[16]

Companhia Brasileira de Propriedades Agrícolas S.A. (BrasilAgro), headquartered in São Paulo, Brazil, is a publicly traded company listed in the Bovespa stock exchange with American depositary receipts (ADRs) traded in the New York Stock Exchange (NYSE). Its 2006 initial public offering (IPO) raised BRL 584 million (about US$286 million) from investors based on a business plan and a promise "to create value by acquiring, developing, and operating properties with sustainable and innovative practices" (BrasilAgro, 2014). The firm did not have any assets and employed only two managers at the time of listing. Since then, it has become one of the leading agricultural land development and farming companies in South America.

The core business of BrasilAgro is the acquisition, development, operation, and sale of rural properties suitable for agricultural production. Once BrasilAgro acquires farmland, it invests in infrastructure, facilities, and technology necessary for efficient agricultural production. It then engages in farming operations aiming to maximize cash flow per area. BrasilAgro selectively divests of a farm when it reaches its optimal value to capture capital gains. The company combines the returns generated from land value appreciation and farming operations, while mitigating production risks with geographic diversification. Its vision is "to be the leading platform for investing in and developing farmland in Brazil" (BrasilAgro, 2014).

With the capital raised in the IPO, BrasilAgro acquired 11 farms in agricultural frontier regions throughout the cerrado. After taking possession of its first farm in July 2007, the firm planted 22,000 ha in its first year of operation. The planted area has increased every year since then, reaching 80,000 ha in the 2013—2014 crop year. In 2014 BrasilAgro had a land portfolio of eight farms with 180,000 ha having an estimated market value of BRL 1.3 billion. Three farms had already been sold, allowing the firm to realize considerable capital gains.

The idea for the formation of BrasilAgro came from a group of investors led by Cresud, a diversified real estate development firm in Argentina with a business unit in farming. Its land portfolio in Argentina consisted of 20 farms totaling 650,000 ha. In the mid-2000s, Cresud was considering global expansion to leverage its land development expertise in other countries. Cresud is currently the controlling shareholder in BrasilAgro, with a 39.7% stake in the company. The remaining shares are traded in the Bovespa stock exchange and are held by minority shareholders. BrasilAgro is listed in Bovespa's New Market, which requires high levels of corporate governance practices and transparency. With the decision to list in the New Market, BrasilAgro was able to raise capital at a competitive cost as it offered more security and transparency to investors.

The board of BrasilAgro is composed of nine directors, of which three are independent. Together, the board of directors and the board of executive officers are responsible for managing BrasilAgro. The board of directors is responsible for establishing long-term strategies and setting general business policies and guidelines. Professional executive officers are delegated responsibility for the day-to-day management of BrasilAgro's business following the resolutions of the board of directors. The board of executive officers is comprised of four professional managers led by Mr. Julio Piza. Julio joined BrasilAgro as CEO in April 2008, when the firm was harvesting its first crop. With formal training in agriculture (bachelor's degree, University of São Paulo) and business (MBA, Columbia University), Mr. Piza had prior professional experience in farm management and as a consultant at McKinsey. The cornerstone of the business model developed by BrasilAgro is its large size, which enables it to benefit from economies of scale. According to Julio, economies of scale are realized in two levels: at the farm level and the corporate level. Farm-level economies of scale include the following: fixed cost dilution (such as overhead expenses and compliance costs with labor, environmental, and tax laws that are exceedingly high in Brazil); ability to attract and retain professional managers and experienced technical staff to run each farm; efficient use of on-farm facilities and infrastructure; and operational efficiencies of modern farm equipment. Economies of scale at the corporate level include commercial advantages in buying farm inputs (e.g., volume discounts) and in negotiating commodity prices or forward contracts (due to higher bargaining power). In addition, its size and access to capital allow BrasilAgro to invest in modern information and communication systems and to develop knowledge to make better commercial and risk management decisions. Perhaps more importantly, size and scale lead to a lower cost of capital and the reduction of price and production risk due to geographic and product diversification.

To benefit from these potential economies of scale, the major challenge of corporate farms like BrasilAgro is that the owners (shareholders) are not involved with farming operations and are distant from the farms. This separation of ownership and management gives rise to conflicts of interest and agency costs[17] between owners, managers, and farm workers. These agency costs are potentially severe in agriculture because of the unpredictable effects of Mother Nature on production (Allen and Lueck, 2002). Julio believes that it is possible for a corporate farm to achieve high performance by means of a well-designed organizational architecture. The organizational architecture of BrasilAgro includes a hierarchical structure with well-defined responsibilities and communication channels between the corporate team in São Paulo and the farm managers, formal control systems, and incentive compensation based on key performance indicators.[18]

Another example of a corporate-style farming entity that was recently formed to develop farmland in the cerrado is Agrifirma Brasil Agropecuária S.A. Agrifirma is a limited company formed in 2008 with initial equity capital provided by RIT Capital Partners and Lord Rothschild – both private equity firms headquartered outside Brazil. In 2011 Agrifirma received another large investment from BRZ Investments, a large Brazilian private equity firm. The current ownership structure of Agrifirma includes two controlling shareholders — Genagro (a holding company owned by the initial investors — RIT and Rothschild) and BRZ — in addition to some minority investors. These are all "passive investors" in the sense that Agrifirma is just one asset in their diversified portfolios. As is the case with private equity firms, they have a limited investment horizon of 7–8 years, after which they expect to exit with considerable capital gains.

As is the case with BrasilAgro, Agrifirma owners (shareholders) are not involved in the business. Each majority shareholder (Genagro and BRZ) appoints two directors to the board, which also includes three seats for independent directors. The board of directors meets on a quarterly basis to set policy and monitor business performance, but board directors are not involved in managing the business. Despite not being a listed company, Agrifirma follows corporate governance rules to ensure transparency and disclosure and to minimize conflicts of interest between shareholders and the senior management team.

The business model adopted by Agrifirma is very similar to the BrasilAgro model described above — to acquire cheap farmland in the Brazilian cerrado, make the necessary investments to bring the land into production, and then maximize cash flow from farming operations. Since 2008, Agrifirma has bought 71,000 ha of farmland in western Bahia, in three clusters of about 24,000 ha each. In 2014 Agrifirma planted 23,200 ha including soybeans, corn, and cotton. In the near future Agrifirma plans to develop and bring into production an additional 27,000 ha in the tree clusters, with an estimated investment of US$1,500 per hectare.

The corporate structure based in São Paulo is comprised of the board of directors and a senior management team of four professionals (CEO, CFO, COO, and legal counsel) assisted by support staff. The firm estimates that this corporate structure costs about US$60 per hectare and serves the purpose of providing a governance structure to attract investor capital at low cost. As the firm grows with the acquisition of more farmland, this corporate cost is expected to be diluted. "Our corporate costs are heavy given the current size of our farming operations. This is why growth is crucial for Agrifirma in the future," according to Fabiano Costa, the CFO.

The COO, Rodrigo Rodrigues, oversees all farming operations conducted in the three clusters in western Bahia. The field staff includes 30 managers responsible for basic administrative functions (e.g., human resources, accounting, finance, commercial, etc.) and 450 farm managers and workers. Each farm cluster is managed by one farm manager who reports directly to the COO. The organizational architecture at the farm level is very similar to BrasilAgro, including a hierarchical structure, formal budgeting and control systems, and performance-based compensation. According to Rodrigo, the major challenge he faces is to recruit, develop, and incentivize human resources. "Agriculture is not a one-man show. You have to develop a good team to execute the business plan laid out by the board of directors. Unfortunately in Brazil, especially in remote frontier regions like western Bahia, talent is a scarce resource. It will take time, patience, and commitment to form a team and an organizational culture focused on delivering results to shareholders."

Agrifirma estimates that overhead costs associated with farming operations are about US$100−120 per hectare. "Organizational architecture is the backbone of any large-scale, corporate farming entity. It is a substitute for the personal ties and informal organization of traditional family farms. However, it takes time to develop; there is a learning curve. The difference is that formal organizational architecture is scalable. Once you make it work, there is no limit to how large you can grow," says Rodrigo.

5.4.3 Economies of Scale

We described in Chapter 1 the impressive productivity gains achieved by Brazilian farmers since the 1970s. Despite this impressive performance, there is a growing efficiency gap between the most efficient and the average farms. In the cerrado, producers operating at the technical frontier − i.e., those who manage their resources most efficiently − enjoyed a total factor productivity (TFP) growth rate of 4.3% per year from 1985 to 2006. The majority of the cerrado producers, however, were not able to match such efficiency gains and thus the average TFP growth rate in the region was only 0.4% (Rada, 2013). This significant productivity gap between farms in the same region suggests considerable room for efficiency improvement.

This growing productivity gap between producers in Brazil also begs the question − what farm-level factors help explain productivity gains? In particular, what is the effect of scale economies on productivity growth in Brazilian agriculture? Unfortunately, there is no definitive empirical evidence to support the claim that large farms are more productive than small farms. One recent study found a non-linear relationship between farm size and technical efficiency among farms in the cerrado, showing that efficiency falls as size rose for farms up to about 1,000–2,000 ha, but beyond this size it starts to increase again (Helfand and Levine, 2004, p. 248). Anecdotal evidence from my field research in Mato Grosso suggests that farmers enjoy economies of scale as they grow up to 2,500 ha. Also, large commercial producers farming more than 10,000 ha appear to benefit from economies of scale. However, producers "stuck in the middle" – not small enough to rely primarily on family labor and not large enough to attract professional management and technical staff – appear to be suffering.

The large-scale-dominated structure of production agriculture in the Brazilian cerrado, and the emergence of corporate farming schemes, is intriguing because the small-scale, family farm production unit has been the dominant form of organization in agriculture since the early days. The literature on farm organization, which focuses on the incentive problems and the resulting inefficiencies when the principal farm operator is not the landowner, provides a convincing economic rationale as to why agriculture remains the last bastion of family organization. A basic finding in this literature is that the separation of ownership and management has an adverse effect on incentives for agents to provide optimal labor effort when output is uncertain. The random and seasonal effects of Mother Nature on agricultural production not only exacerbate agency costs in corporate farms, but also limit the gains to labor specialization that can occur in factory-style or hierarchical organizations. As a result, evolution to corporate-style farms is limited, allowing small family farms to dominate and persist, particularly in broadacre farming that is more influenced by random factors such as weather and pests. Given this context, the main issue is – will these experiments in corporate-style agriculture in the Brazilian cerrado region survive?

5.5 PRODUCER ORGANIZATIONS

Agriculture in the cerrado is very dynamic and no dominant form of organization has yet emerged. As explained above, two types of commercial farms appear to be competitive in the agricultural frontier – family and corporate farms. The latter includes very large farming operations controlled by family groups (such as Produzir S.A. and Grupo Bom Futuro) or corporate entities (such as BrasilAgro and Agrifirma). Economies of scale, access to capital, and professional management are their main sources of competitive advantage. The former includes family producers farming up to 2,500 ha. They have sufficient scale to be able to make the necessary investments in technology and machinery to achieve high productivity levels and reduce per-unit production costs. Relative to large family groups and corporate farms, independent family farmers face higher input costs, are more dependent on farm input suppliers and traders to obtain credit, and have disadvantages in accessing markets. As a result, these commercial family farms increasing rely on producer associations and cooperatives to remain competitive.

5.5.1 The Soybean and Corn Producer Association (Aprosoja)

Aprosoja was created in 2005 as a response to the crisis that affected yields, reduced margins, and caused widespread financial problems among farmers. The inspiration for the formation of Aprosoja came from two successful producer organizations in the state – FMT, which we described above, and AMPA, the cotton producers' association. Soybean producers from across the state provided support to a mandatory check-off program, called *Fundo de Apoio à Cultura da Soja* (FACS), to fund projects that would benefit soybean production in the state. The FACS was instituted by a state law in December 2005, and the mandatory contribution by producers was initially set at BRL 5 cents per soybean bag. This mandatory producer contribution was matched by state funds. FACS funding reached BRL 36 million in 2014 with oversight from a board of trustees nominated by the state government and two producer associations (Aprosoja and Famato). Producer organizations like Aprosoja and FMT and public research institutes such as EMBRAPA and the Mato Grosso Agricultural Economics Institute (IMEA) may submit proposals to the FACS board to fund projects that will generate public goods to soybean producers.

All soybean producers in Mato Grosso must contribute to the FACS, but membership in Aprosoja is voluntary. Aprosoja had 5,412 producer-members in 2014, with 75% of them farming less than 2,000 ha. The association is organized by region – 22 regions with at least 40,000 ha in soybean production are represented in the council of delegates. Each region nominates two to four delegates to the council based on soybean-planted area. The council of delegates elects the board of directors and the fiscal board. All delegates and board directors must be active soybean producers, and they do not receive remuneration to serve at Aprosoja. The board of directors is in charge of managing the association with the assistance of a full-time, professional executive director. Aprosoja had a staff of 38 professionals and a budget of BRL 18 million in 2014.

The mission of Aprosoja is to "enable the competitiveness and sustainability of soybean and corn producers in Mato Grosso." To achieve this mission, the association provides several services to members: information, education, advisory, and political representation. Its activities are coordinated by five committees: agricultural defense, logistics, farm management, agricultural policy, and sustainability. Each committee is coordinated by a producer and managed by technical staff. The agricultural defense, logistics, and agricultural policy committees are directly involved with the state and federal governments to inform policies affecting soybean producers. The other two committees — farm management and sustainability — manage programs providing services directly to members.

One program that directly affects producers is called *Soja Plus*. The program was initiated in 2012 with two main objectives: to enhance the economic, social, and environmental management of soybean farms and to help farmers comply with labor and environmental laws. To participate in the program, the producer must first attend a training program. Then his farm is evaluated by Aprosoja field staff according to 143 performance indicators. The objective of this evaluation is to identify opportunities for improvement and to assist the farmer in adopting labor and environmental practices to comply with the law. The ultimate goal is to equip the producer with the tools and assistance to engage in continuous improvement of farm practices and thus increase economic, social, and environmental performance. In 2014, the field staff visited, evaluated, and provided advice to 620 producers and the 2015 goal is to reach out to 1,000 producers.

Another program that also helps producers increase performance is called *Projeto Referência*. The program is based on a web-based benchmarking application that allows soybean and corn producers to compare production and financial performance against peers in the state. In 2014, 120 producers that planted 185,000 ha of soybean voluntarily opted to participate in the project. According to Cid Sanches, the manager in charge of *Projeto Referência*, Aprosoja does not charge a fee for this service, but the benchmarking application requires that the producer have production and accounting information available and be willing to upload this information in the system. "We protect the privacy of the information but most producers in the state do not keep detailed records or are not willing to spend the time to effectively participate in the project."

Another program aimed at producers focuses on providing information and support for collective action. Results from *Projeto Referência* indicate that farmers engaged in collective action — e.g., co-investing in grain storage facilities or buying farm inputs together — have lower production expenses and are able to market production with higher prices. Mr. Piccoli, who is the producer coordinating this program, believes that "collective action is the only means that small and middle-sized producers can compete on the same level-playing field as large family groups and corporate farms. It is the only way for them to survive as commercial producers in Mato Grosso."

5.5.2 New-Generation Cooperatives

We saw above that private colonization organizations played a critical role in providing the initial conditions for families to migrate from the south to the agricultural frontier. However, poor transportation infrastructure, distance to markets, insufficient grain storage, and high transaction costs to obtain credit and farm inputs were common challenges of the early colonists in the 1970s and 1980s. They decided to organize agricultural cooperatives to overcome these market failures and provide the missing services they needed to be successful.

One example of this first wave of cooperatives in Mato Grosso is Coopercana in the eastern part of the state. Founded in 1975, Coopercana played a central economic role in regional development as follows. First, it invested in warehouses and grain storage facilities across the region. The cooperative stored the grain and marketed the production to the government under the minimum price program. In doing so, it enabled farmers to have access to governmental subsidies, which they otherwise would not have. The cooperative also assisted the farmers in accessing subsidized credit from Banco do Brasil. The state bank funneled the total value of loans to the cooperative, which then redistributed the funds to the farmers. Coopercana also provided farmers with agricultural inputs and agronomic services. Inputs were supplied at a discount for cooperative members, which reduced the need for farmers to travel to Barra do Garças, the main local market, located 300 km away from the settlements. Coopercana also disseminated information about suitable agricultural practices for production on poor-quality cerrado soils.

In the early 1980s, Coopercana was instrumental in helping farmers improve soil fertility and thus decrease dependence on rice with production diversification. Based on agronomic recommendations from EMBRAPA, the cooperative provided lime, chemical fertilizers, and technical advice to encourage farmers to invest in soil fertility and avoid soil degradation. With improved soil fertility, the cooperative introduced new crops to the region. In collaboration with EMBRAPA, Coopercana implemented crop experiments and provided important agronomic information to farmers. The cooperative eventually tested 16 tropical soybean cultivars and other crops, including corn, sorghum, beans, coffee, manioc, garlic, and forage in its experimental fields. In 1987 the cooperative introduced its own soybean variety, called Canarana. New technologies were disseminated to farmers in field days and educational programs. With the availability of adapted cultivars and credit for soil improvement, agricultural production and productivity in the region increased dramatically. Between 1983 and 1993, soybean production increased from 5,000 to 40,000 ha, while corn production increased from 700 to 3,000 ha.

Production growth and agricultural development in eastern Mato Grosso provided the impetus for the growth of Coopercana in the 1980s. In addition to the services provided to farmers explained above, the cooperative diversified and engaged in several vertical integration projects upstream and downstream in the value chain. These growth projects included: lime processing and marketing; corn, rice, and soybean seed production and marketing with its own brand name; a network of supermarkets and gas service stations; and a livestock processing plant.

Coopercana was not an isolated case. Several multipurpose, local cooperatives were formed in the 1970s and 1980s across Mato Grosso by the early pioneers. Because these pioneers came mostly from the southern region, the cooperatives they formed followed the traditional model of the cooperatives they knew or were members of in southern Brazil. This first wave of cooperative development in the cerrado was also influenced by increased federal intervention in cooperatives that lasted until 1988. With the enactment of the 1971 cooperative law, the federal government reserved the right to oversee the organization and functioning of all types of cooperatives in the country. Between 1966 and 1988, a state agency known as INCRA (*Instituto Nacional de Colonização e Reforma Agrária*) regulated and controlled agricultural cooperatives. The 1971 law also defined the legal status of cooperatives and set rules for their formation and functioning.

Coopercana and most of the first-wave cooperatives in Mato Grosso went bankrupt in the early 1990s. There were several factors that led to their demise. With unsustainable debt levels, these cooperatives could not survive the period of hyperinflation of the late 1980s and the economic liberalization and macroeconomic reforms introduced in the early 1990s. Farmers, in turn, also suffered from the economic crisis and the end of agricultural subsidies, and many were forced to exit agriculture. In addition, there were some serious management and governance issues that prevented these cooperatives from taking the necessary steps to restructure and survive the crisis.

With the failure of most first-wave cooperatives in the 1990s, market failures persisted in Mato Grosso despite the growth of agricultural production. A subset of the commercial farmers aggressively pursued horizontal expansion to benefit from economies of scale in production and have more bargaining power to buy farm inputs and to market agricultural commodities. These farmers also invested heavily in on-farm infrastructure, especially in modern machinery and grain storage facilities. The small and medium-sized commercial producers, however, faced increasing market failures and transaction costs. Since the mid-1990s, these farmers have overcome collective action challenges and formed 43 new-generation cooperatives across the state. In 2014, I conducted a survey research of 30 of these 43 cooperatives with support from Mato Grosso Cooperative Organization (OCB/MT) and Aprosoja staff. Personal interviews were conducted with 11 board chairs, seven CEOs, and 12 senior managers of these new-generation cooperatives. The interviews were complemented with information from annual financial reports and incorporation statutes and bylaws provided by the cooperatives.

The average cooperative in the sample had 10 years of operation in 2014, and only five cooperatives were formed before 2000. Despite being relatively new, these cooperatives combined had 1,262 full-time employees and 1,546 producer-members in 2013. The total land in farms of cooperative members added up to 2.4 million hectares. According to our estimates, these cooperative members planted 20% of the soybean and corn area and 90% of the cotton area in Mato Grosso in 2013. The number of members averaged 55 per cooperative, ranging from 20 to 197. The members of the new-generation cooperatives were largely small and medium-sized commercial producers, farming on average 1,000 ha in the eastern and central-north regions and 1,500 ha in the western region of the state.

The great majority of the new-generation cooperatives were formed for "defensive" purposes with the objective of protecting margins at the farm level.[19] They provide missing services to members – in particular, grain storage, cotton ginning, and cotton quality classification – and organize input supply and commodity marketing pools to have more bargaining power when dealing with farm input suppliers and commodity traders. Only two of the new-generation cooperatives were formed with the original purpose of adding value to farm commodities – one in sugarcane processing and the other in biofuels. Of the 28 cooperatives formed to play defense, three have subsequently invested in value-added projects. For example, Cooperfibra, the largest cooperative in the state, invested BRL 45 million in 2011 in a state-of-the-art cotton-spinning plant with capacity to produce 1,200 tons of cotton yarn per month.

The organizational characteristics of these new-generation cooperatives are strikingly different from the traditional model adopted by the first wave of cooperatives in the state and their counterparts in the southern region. The first major difference is that they adopt a selective membership policy rather than open membership. In other words, these cooperatives implement stringent criteria for entry of new members. New members have to be introduced by an existing member, and their financial condition and reputation are scrutinized by the board of directors. As a result of the failures of most first-wave cooperatives, farmers are risk-averse and do not want to bear the risks of doing business with other farmers they cannot trust.

The second difference is that the new-generation cooperatives are conceived as "extensions of the farm," not as a business enterprise. They exist to provide services to members and to increase their profit margins, not to generate net income (profits). One of the cooperative leaders explained his cooperative philosophy as follows. "In the southern region, the cooperatives are strong, but the farmers are weak. Here in Mato Grosso, the farmers are strong and the cooperative is just a means for us to achieve our goals." This philosophy means these cooperatives closely follow the "service at cost" principle. For example, instead of purchasing farm inputs and then selling them with a margin to members, the new-generation cooperatives organize farm input pools. The cooperative first pools the farm input needs of members – for fertilizers, chemicals, and seeds – and then bargain with farm input providers for the best price. The actual transaction occurs directly between the input supplier and each member and thus savings are passed directly to members. A similar pooling arrangement occurs for commodity marketing of soybeans, corn, and cotton. In this case, it is beneficial for the farmer to market through the cooperative for tax reasons. The cooperative transfers the full price received by the buyer to each member. In both the farm input and commodity marketing pools, the cooperative charges a service fee to act on behalf of the member. In doing so, most of the cooperatives do not generate significant net income at the end of the fiscal year as savings are distributed directly to members as better prices.

The third distinguishing feature of these new-generation cooperatives is that members are required to invest upfront and commit to cover operating costs in proportion to usage whenever an investment is made in fixed assets. For example, some members have on-farm storage but others do not. In this case, the cooperative organizes a separate capital pool for the group of farmers who want to invest in cooperative storage. Each farmer actually invests in a share of the grain storage facility and commits to cover his respective share of the facility operating costs on an annual basis. A similar model is followed to invest in value-added processing ventures. Cooperfibra has 197 producer-members, but only 32 of them decided to invest in the cotton-spinning plant. The investment was organized as a separate capital pool under the cooperative structure with upfront investment from the 32 member-investors. Profits (and losses) are distributed in proportion to the equity stake of each member-investor in the processing plant.

With the exception of three cooperatives that are in their start-up phase, with less than 3 years of operation, the other 27 cooperatives in the sample are in a growth phase. The total revenue generated by the cooperatives increased from BRL 1.8 billion in 2009 to BRL 4.5 billion in 2013. On average, total revenue per cooperative more than doubled, from BRL 74 million to BRL 166 million, in the last 5 years. Most of this growth has been horizontal, with selective entry of new members and more business volume generated per member. The bundle of services provided to members does not appear to increase in the growth period as most cooperatives do not diverge from their original purpose. As explained above, only three of the new-generation cooperatives have invested in downstream, value-added projects.

5.6 SUMMARY

In this chapter we continued to explore the economics and organization of Brazilian agriculture with a focus on how the cerrado has been developed since the 1970s and has become a breadbasket since the 2000s. Given the diversity of agriculture in the cerrado, a biome that covers 200 million hectares, this chapter described how agriculture developed in the state of Mato Grosso. Crop production increased from 3 to 48 million tons between 1977 and 2014, while the number of beef cattle soared from 3 to 28 million head, making Mato Grosso the largest crop and beef cattle producer in the country.

This chapter explained how policies adopted by the military regime in the 1970s provided the initial impetus for agricultural development in the cerrado. These policies, however, were short-lived and insufficient to explain how commercial agriculture thrived in the region and achieved international competitiveness despite large distance to ports, poor infrastructure, and market failures of all sorts. This chapter explained and provided examples of public–private partnerships that developed technologies adapted to the cerrado conditions and made them available to farmers. It also described how private colonization firms and agricultural cooperatives played a crucial role in settling pioneers who came to Mato Grosso from the southern region in the 1970s and 1980s and provided them with the technologies and services necessary to farm in the agricultural frontier.

In addition to ill-functioning markets and poor infrastructure, the early pioneers had to survive several crises in the late 1980s, early 1990s, and again in the early 2000s. This chapter discussed and provided examples of how they were able to survive these crises and become commercial producers in the agricultural frontier. In addition to good fortune, becoming a successful, commercial farmer in the frontier required entrepreneurship, adoption of modern technologies, economies of scale, and the organization of producer associations and new-generation cooperatives. These producer organizations provide critical services and collective goods to producers, counter the market power of farm input providers and commodity traders, and represent producers' interests in policy making. The commercial producers who were able to survive these crises benefited from favorable marketing margins between 2006 and 2014 and continued to expand production with increased levels of productivity.

In contrast to the southern and southeastern regions, there is no clear dominant organizational form in the Brazilian cerrado. Two alternative organizational forms appear to be emerging – small and medium-sized commercial producers with a traditional family farm organization and very large family groups and corporate-style farming entities with hierarchical organizations, professional management, and access to outside capital. The producer organizations and new-generation cooperatives formed since the early 2000s in Mato Grosso provide a level playing field for family farmers to compete with large family groups and corporate farming entities that are consolidating farmland in the cerrado. Only time will tell whether they will be as successful as their counterparts in the southern region in linking farmers to markets.

NOTES

1. About 28% of the principal operators of farm establishments in Mato Grosso in 2006 were born in the southern region and 24% in the southeastern region. Only 22% of these farm operators were born in Mato Grosso (IBGE, 2006).

2. In order to implement its development policy and integrate the Amazonian region with the rest of the country, the Brazilian military government created the Legal Amazon, a region that includes the states of Acre, Amapá, Amazonas, Pará, Roraima, Mato Grosso, Tocantins, and Maranhão west of the 44th meridian. The Legal Amazon thus includes three biomes: the Amazon rainforest, part of the cerrado, and part of the Pantanal wetlands.

3. In 1968 the road system was entirely federal. Over time, the states also contributed to road building, which reached 37,000 km in 2001. Most paved roads, however, are found in the federal component of the system (Walker and Reis, 2007).

4. This was accomplished with the *Programa de Polos Agropecuários e Agrominerais da Amazônia*, designed to attract investment with subsidies and tax relief (Hecht et al., 1988).

5. Jepson (2006a,b) provides detailed accounts of the role of private colonization projects in the Legal Amazon. She concludes that "state incentives were, in fact, insufficient to *cause* the frontier expansion. Central to Brazil's frontier historical geography are private colonization cooperatives and firms, both of which developed into critical organizations in the process of agricultural expansion" (Jepson, 2006a, p. 858).

6. According to industry estimates, the economic losses caused by Asian rust disease amounted to US$18 billion between 2002 and 2010, including yield losses and increased production costs associated with disease control.

7. Under the same growing conditions, soybean cultivars with a longer cycle (125–130 days) are more productive than cultivars with a shorter cycle (less than 110 days), as they have more time to develop their root system and leaf area, which increases nutrient accumulation in the plant and subsequently in the grains. Thus, early planting of soybean cultivars with shorter cycles does not allow the crop to express the genetic potential of the seed (Specht et al., 2014).

8. Dario Hiromoto was born in 1962 in Londrina, PR, the son of immigrants from Japan who settled in Marilia, SP. He obtained his bachelor's degree in agronomy from the College of Agriculture of the University of São Paulo (ESALQ/USP) in 1986 and subsequently his master's and Ph.D. degrees in genetics and plant breeding from the same university. His early passing in 2009 was a major setback for agricultural development in the cerrado.

9. We explained the pivotal role that Dr. Kiihl played in the "tropicalization" of soybean cultivars in Chapter 2. He is considered the "father of soybeans" in Brazil.

10. Agrosilvopastoral systems are land use systems of production and conservation based on forestry practices complementary to pre-existing agricultural and livestock activities. ASP systems may be applied in a wide range of ecological and productive conditions. For more details, see Russo (1996).

11. The team of researchers at EMBRAPA Agrossilvipastoril have disciplinary backgrounds in the following fields: integrated pest management, plant diseases and weeds, nematology, plant science, chemistry, soil physics and biology, water resources, carbon dynamics, greenhouse gases, climate change, agro-meteorology, precision agriculture,

remote sensing, forest management and restoration, genetic resources, bioenergy, biomass, animal health, animal production, pasture management, ethnobiology, horticulture, aquaculture, economics, and regional development.

12. The exchange unit used by farmers in Mato Grosso is not the national currency, the real, but soybean bags. One soybean bag weighs 60 kg.

13. The market rental price of farmland in Sorriso, MT, was 8—11 soybean bags per hectare in 2014.

14. C-Vale is the fourth-largest agricultural cooperative in Brazil (see Table 3.6 in Chapter 3).

15. I am not adopting the Brazilian definition of "family farm," which is based on farm size and the employment of non-family labor, thereby excluding the majority of commercial producers. Rather, I follow the US Department of Agriculture definition, which defines a family farm when the principal operator of the farm business is the landowner irrespective of farm size (see Hoppe, 2014).

16. The agricultural production areas in the Mapitoba region are on average 1,000 km closer to the ports than in Mato Grosso, with a 30% advantage in transportation costs.

17. As is well established in the economics literature, agency costs arise when decision agents do not bear the wealth effects of their decisions (Jensen and Meckling, 1976). These agency costs include the costs of monitoring managers and the costs resulting from managerial opportunism (Hansmann, 1996).

18. Refer to Chaddad (2014) for a detailed analysis of BrasilAgro's business model and organizational architecture.

19. Some cooperatives, however, are formed to play "offense" with the intent of adding value to agricultural commodities (Cook, 1995).

REFERENCES

Allen, D.W., Lueck, D., 2002. The Nature of the Farm: Contracts, Risk, and Organization in Agriculture. MIT Press, Cambridge, MA.

BrasilAgro, 2014. Mission and Vision Statements. Available from: <www.brasilagro.com.br> (accessed 07.08.14.).

Chaddad, F.R., 2014. BrasilAgro: organizational architecture for a high performance farming corporation. Am. J. Agric. Econ. 96 (2), 578—588.

Companhia Nacional de Abastecimento — CONAB, 2015. Séries Históricas de Área Plantada, Produtividade e Produção, Relativas às Safras 1976—77 a 2014—15 de Grãos. Available from <www.conab.gov.br> (accessed 22.02.15.).

Cook, M.L., 1995. The future of U.S. agricultural cooperatives: a neo-institutional approach. Am. J. Agric. Econ. 77 (5), 1153—1159.

Fundação de Apoio à Pesquisa Agropecuária de Mato Grosso, 2010. Dario Hiromoto: O Legado de um Semeador. DBA Dórea, São Paulo, SP.

Hansmann, H., 1996. The Ownership of Enterprise. The Belknap Press of Harvard University Press, Cambridge.

Hecht, S.B., Norgaard, R.B., Possio, G., 1988. The economics of cattle ranching in eastern Amazonia. Interciencia 13, 233—240.

Helfand, S., Levine, E., 2004. Farm size and the determinants of productive efficiency in the Brazilian Center-West. Agric. Econ. 31, 241—249.

Hoppe, R.A., 2014. Structure and finances of U.S. farms: family farm report, 2014 edition. Economic Information Bulletin 132. Economic Research Service, U.S. Department of Agriculture, Washington, DC. Available from: <http://www.ers.usda. gov/media/1728096/eib-132.pdf> (accessed 25.02.15.).

Instituto Brasileiro de Geografia e Estatística — IBGE, 2006. Censo Agropecuário, Brasilia, DF. Available from: <www.ibge.gov.br> (accessed 09.11.14.).

Jensen, M.C., Meckling, W.H., 1976. Theory of the firm: managerial behavior, agency costs and ownership structure. J. Financ. Econ. 3 (4), 305—360.

Jepson, W., 2006a. Private agricultural colonization on a Brazilian frontier, 1970—1980. J. Hist. Geogr. 32 (4), 839—863.

Jepson, W., 2006b. Producing a modern agricultural frontier: firms and cooperatives in Eastern Mato Grosso, Brazil. Econ. Geogr. 82 (3), 289—316.

Klink, C.A., Machado, R.B., 2005. Conservation of the Brazilian cerrado. Conserv. Biol. 19 (3), 707—713.

Mueller, C. C., 2003. Expansion and Modernization of Agriculture in the Cerrado: The Case of Soybeans in Brazil's Center-West, Working Paper 306, Department of Economics, Universidade de Brasilia, Brasilia, Brazil.

Nassar, A.M., 1998. Fundação MT: Um Caso de Ação Coletiva no Agribusiness, Case study prepared for the VIII International Agribusiness Seminar, PENSA, University of São Paulo. Available from: <pensa.org.br> (accessed 23.02.15.).

National Agricultural Statistics Service — NASS, 2015. National Statistics for Soybeans. Available from: <www.nass.usda.gov> (accessed 15.02.15.).

Rada, N., 2013. Assessing Brazil's cerrado agricultural miracle. Food Policy 38, 146—155.

Russo, R.O., 1996. Agrosilvopastoral systems: a practical approach toward sustainable agriculture. J. Sustain. Agr. 7 (4), 5—16.

Specht, J.E., Diers, B.W., Nelson, R.L., Toledo, J.F.F., Torrion, J.A., Grassini, P., 2014. Soybean Yield Gains in Major U.S. Field Crops. CSAA, Madison, WI, CSSA Special Publication 33, pp. 311—355.

Walker, R., and Reis, E., 2007. A Basin-Scale Econometric Model for Projecting Future Amazonian Landscapes, Final Report, NASA, Greenbelt, MD.

Walker, R., DeFries, R., Vera-Diaz, M.C., Shimabukuro, Y., Venturieri, A., 2009. The expansion of intensive agriculture and ranching in Brazilian Amazonia. Amazonia and Global Change, Geophysical Monograph Series 186. American Geophysical Union, pp. 61—81.

CHAPTER 6

Conclusions

Contents

6.1 SUMMARY OF MAIN FINDINGS

It is the early 1970s. Brazil is enjoying an "economic miracle" with high GDP growth rate, industrialization, and urbanization. Yet its agricultural sector lags behind. Land tenure is concentrated, with very large, unproductive farms called *latifúndios*. Smallholders use traditional production systems with low input use and labor-intensive techniques. Rural poverty is a major national problem and millions of rural poor migrate to *favelas* in big cities. Commercial farms are relatively few and depend on imported cultivars, farm inputs, and machinery. Economic policies of the import substitution model do not provide incentives for agricultural development. Farm productivity is low; the country is a net importer of food and a recipient of food aid from abroad. The military government adopts policies to increase agricultural production, decrease food prices, and provide food security to an increasingly urban population. These policies include public investments in agricultural R&D and extension, price controls, and subsidized rural credit. Additional policies target specific sectors or regions, such as the *Pro-alcohol* program and the National Integration Plan (PIN) to integrate the Amazon region to the rest of the country.

Forty years later, in the early 2010s, Brazil is a top-five producer of 36 commodities globally and the largest agricultural *net* exporter in the world. Domestic food prices are 80% lower in real terms compared to the early 1970s. Brazilian families spend only 16% of the household budget on food, which is equivalent to their counterparts in developed countries. Agricultural production grew fourfold between 1975 and 2010,

F. Chaddad: *The Economics and Organization of Brazilian Agriculture.*
DOI: http://dx.doi.org/10.1016/B978-0-12-801695-4.00006-9
147

bolstered by productivity gains of 3% per year. The food and agricultural sector in Brazil was transformed from a traditional to a global model.

So, if you take a poor, tropical country with low productivity levels in agriculture and challenged by food security, high food prices, and rural poverty, all you need to do is to adopt the right set of policies and invest in agricultural research and extension. Right? This book attempted to show that the story is more complex than that. Yes, availability of natural resources matters. Yes, you need to invest in the development of agricultural technologies adapted to the local conditions. And, yes, you need to provide subsidies and incentives for farmers to adopt these technologies and increase agricultural productivity. In this book we called these factors *enabling conditions*.

However, we hypothesized that *in addition to individual characteristics (and luck), success of a farmer depends on access to basic factors of production (land, technology, and credit) and access to markets.* Embedded in this hypothesis are two factors associated with agricultural development that are often overlooked — entrepreneurship and value chain organization. This book provided several examples of farmers (Eugênio Pinesso in Chapter 1, Nelcon Piccoli, Silvésio de Oliveira, and Eraí Maggi in Chapter 5), who left southern Brazil to try their luck in the cerrado in the 1970s and 1980s. These entrepreneurs had the guts to leave it all behind and relocate to a remote, frontier region. They had the foresight and capacity to adopt modern technologies, were resilient to several economic crises, and adapted to constant changes in the institutional environment. They were all peasants or smallholders in the south who became successful commercial farmers in the cerrado. This book also presented examples of entrepreneurs who stayed in southern Brazil and helped transform their firms and industries into world leaders (Chapters 3 and 4). These entrepreneurs include Dutch and German colonists who arrived in Paraná in the 1950s and formed successful cooperatives, such as Castolanda and Agrária, that to this day provide services and market access to family farmers. In São Paulo, sugarcane producers formed Copersucar and hired leaders such as Luís Pogetti to transform their cooperative into the largest sugar and ethanol trader in the world. The case study of Cosan showed how a single-plant, family-owned sugarcane processor in the 1980s consolidated the industry, listed shares in the stock exchange, and became one of the largest conglomerates in the country. We also described how the Cutrale and Fisher (Citrosuco) families evolved from orange growers in the 1970s to multinational orange juice processors and exporters.

Obviously, these entrepreneurs benefited from the policies, subsidies, and protection provided by the Brazilian government in the 1970s and 1980s. But, since the early 1990s, they have survived and prospered in a deregulated and liberal market environment, without safety nets and prone to several economic crises. This book presented several examples of farm-, firm-, and industry-level changes adopted by these entrepreneurs to increase productivity and build world-class organizations. It is hard to imagine how Brazil would have become a powerhouse in agriculture without the entrepreneurship of these individuals and families. Many farmers and entrepreneurs, however, did not make it through these crises and industries consolidated as a result. As I write these concluding thoughts in early 2015, world commodity prices are softening and Brazil is going through a deep recession coupled with high inflation and political instability. I suspect that we will see more structural changes and consolidation in the near future.

Another factor often overlooked by scholars interested in agricultural development is the *central* role of value chain organization. How do technology, credit, and modern farm inputs reach farmers? How does farm production reach consumers in domestic and export markets? Well, you need to organize value chains linking the successive stages of technology development, farm input supply, agricultural production on the farms, and the post-farm-gate stages of aggregation, transportation, processing, marketing, and distribution. In addition, the participants in each of the stages along the value chain must be coordinated to provide an affordable, safe, and quality food product to the end consumer. This is no small feat. This book presented several examples of value chain organization in the main agricultural regions in Brazil — cooperatives and contract farming arrangements in the southern region (Chapter 3) and vertically integrated agribusiness in the southeastern region (Chapter 4). Chapter 5 showed the recent emergence of family groups, corporate farming entities, and new-generation cooperatives in the cerrado, but the organization of value chains is still a work in progress in the agricultural frontier.

These well-organized and coordinated value chains are complemented by several types of producer organizations that provide public goods to value chain participants. Chapter 3 discussed the role of OCEPAR in monitoring the functioning of and providing educational services to farmer-owned cooperatives in Paraná. In addition, OCEPAR created a research department in 1974 that was subsequently reorganized as a federated cooperative (Coodetec) that invested in germplasm development and plant breeding programs to provide seeds to farmers via cooperatives. Paraná producers formed research foundations such as Fundação ABC and FAPA to develop production systems adapted to local conditions and provide outreach and extension services to family farmers. Chapter 4 explained how producer organizations such as UNICA, CTC, and Consecana provide policy support to the sugarcane sector, open new markets, introduce sustainable practices, develop new technologies, and harmonize the relationship between sugarcane growers and processors. Similarly to their counterparts in the traditional regions, cerrado farmers developed organizations such as Aprosoja, AMPA, and the Mato Grosso Research Foundation (FMT) to provide political representation, business development services, agricultural technology development, and extension to farmers in the agricultural frontier (Chapter 5).

Interestingly, most of these producer organizations were developed by farmers and are governed by farmers. In addition to entrepreneurship at the farm level, producers were able to overcome collective action challenges to design, fund, and govern a myriad of producer organizations. It is hard to imagine how agriculture would have developed in Paraná, São Paulo, and Mato Grosso without cooperatives, research foundations, and producer associations. They provide the institutional and organizational glue for value chain development and coordination. They reduce transaction costs, provide missing services, and countervail market power. After going through all these case studies, vignettes, and stories, it seems to me that both individual and collective entrepreneurship of farmers are essential to agricultural development in Brazil.

6.2 A COUNTERFACTUAL EXAMPLE

The aim of this book was not to develop new theory, nor to test existing theories. Rather, it was to call attention to two oft-neglected factors associated with agricultural development — entrepreneurship and value chain organization. If you are still not convinced that these two factors matter, please consider the following counterfactual example. Despite an abundance of natural resources, public investments in agricultural R&D and public policy support, Brazil continues to be a net *importer* of milk and dairy products. Why is this so?

First, because the milk sector in Brazil is still protected with import tariffs and quotas. Since the formation of the Mercosur trade block in 1991, dairy products imported from non-Mercosur countries face a 28%

import tariff. In the early 2000s, the Brazilian government added anti-dumping tariffs for dairy products imported from the European Union and New Zealand. More recently, Brazil introduced import quotas for milk powder from Argentina in 2008. As a result of such protectionist policies, Brazilian milk producers are shielded from imports and milk prices received at the farm gate are significantly higher than in more competitive countries. Import restrictions and relatively high producer prices dull incentives for efficiency improvements and productivity gains at the farm level. Not surprisingly, Brazilian milk producers have the lowest productivity levels among the top-20 countries, with the exception of India. While 35,000 farms are required to produce 1 billion liters of milk in Brazil, the same production is achieved by less than 600 farms in efficient countries like New Zealand and the United States. Simply put – no entrepreneurship, no productivity gains, no international competitiveness.

The second major reason why Brazil continues to have low productivity levels in milk production and is a net importer of dairy products is that the value chain is not well organized. Historically, cooperatives played an important role in organizing value chains for milk and dairy products. Between the 1930s and the mid-1960s, local dairy cooperatives were formed in the emerging milksheds in southern and southeastern Brazil. At that time, Brazilian milk producers faced market failures in input and output markets and, therefore, formed dairy cooperatives to supply farm inputs at affordable prices, provide missing services (such as credit and technical assistance), to countervail the market power of middlemen and to facilitate access to urban markets. As these local cooperatives developed and the Brazilian population became increasingly urban, local cooperatives got together to invest in regional cooperatives to process and add value to milk collected from their producers. Regional dairy cooperatives were thus organized in several states including Rio Grande do Sul (CCGL), Paraná (CCLP), São Paulo (CCL), and Minas Gerais (CCPR). These cooperative dairy processors competed with private companies in large cities such as Belo Horizonte, Curitiba, Porto Alegre, Rio de Janeiro, and São Paulo.

With subsidies, protection, and industry regulation, the decades of the 1960s through the 1980s provided a positive environment for dairy cooperative development in Brazil. Cooperatives achieved 70% national market share in milk procurement by the end of the 1980s. With price floors at the farm gate and high import barriers, milk production expanded in the country. Local cooperatives benefited from output growth due to their proximity to members and capillarity of their milk collection systems. In addition, regional cooperatives invested in dairy processing facilities and developed distribution channels for branded dairy products. During this growth period, regional cooperatives became dominant players in the major dairy regions. Cooperative brand names – such as *Elegé* in Rio Grande do Sul, *Paulista* in São Paulo, *Batavo* in Paraná, and *Itambé* in Minas Gerais – were very prominent among urban consumers. Of the top 15 dairy companies in 1989, six were cooperatives.

The business environment changed radically in the 1990s. Dairy markets were completely deregulated, as the government discontinued its price control programs. Fostered by rising incomes and urbanization, multinational food processors and retailers entered or increased their investments in Brazil during the 1990s. Rivalry in the dairy industry was particularly affected by the entry of Parmalat in the early 1990s with the acquisition of domestic companies and new product introductions. As consumers started to put an increasing premium on quality, safety, and convenience of food products, UHT milk consumption increased significantly. UHT milk consumption soared from 200 million liters (less than 5% of total fluid milk consumption) in 1990 to 5 billion liters (or 75% of total fluid milk consumption) in 2005. Lastly, the dairy industry was also significantly affected with the introduction of Normative Instruction 51 (IN 51) in the late 1990s – a federal regulation that required that milk be collected from refrigerated tanks (or coolers) on farms. IN 51 had the objective of increasing milk quality and safety, but its impact on dairy cooperatives was negative. IN 51 required dairy processors to collect milk from producers with refrigerated trucks, which had scale effects and undermined the competitive advantage of local cooperatives. With the advent of refrigerated milk collection, dairy processors could collect milk from more distant milksheds and bypass the traditional procurement system of local cooperatives.

As a result of industry deregulation, technological change, competition from multinational companies, and increasing bargaining power of retailers, the market share of dairy cooperatives in Brazil fell from 70% in the early 1990s to 40% of total milk procurement in 2002. The share of cooperatives in dairy processing also decreased during this period because many cooperative processing plants and brand names were sold to multinational corporations. About 70% of the milk collected by cooperatives was processed in cooperative plants, while the remainder was sold in spot markets to private processors. In the early 2000s, only two cooperatives were found among the top ten dairy processors.

Cooperatives lost competitiveness and market share, but no alternative value chain organization emerged to provide technical assistance, credit, and farm inputs to farmers and to effectively connect them to markets, with the exception of a few cooperatives like Castrolanda in Paraná (see Chapter 3). The industry structure is fragmented, with more than 1.2 million milk producers and 2,000 processors operating in local or regional markets. Private processors have arms-length, transactional relationships with their farmer-suppliers. No contract farming arrangements exist to enable effective vertical coordination between producers and processors, which is the case in the poultry and pork industries (Chapter 3). With the lack of effective producer organizations and well-coordinated value chains, it is not surprising that 30% of milk production is informal (i.e., milk is marketed without federal inspection), milk quality is low, there are constant frauds in the sector, and productivity is below par. Unlike Brazil, the major milk exporters in the world — New Zealand, the United States, and northern European countries — have well-organized value chains dominated by farmer-owned cooperatives and producer organizations.

6.3 CHALLENGES

Sustainable development requires growth that is inclusive and broadly shared, as well as environmentally sound. Brazilian agriculture has been transformed since the 1970s with increasing levels of productivity and well-coordinated value chains. It has solved food security concerns, while reducing domestic food prices and generating increasing surpluses exported to international markets. Notwithstanding the positive economic impacts associated with increased production and productivity gains in Brazilian agriculture, some challenges remain. These challenges include environmental and social issues. In what follows, the chapter describes progress being made to ameliorate environmental and social concerns associated with agricultural development in Brazil.

6.3.1 Environmental Issues

The major environmental concern is that the development of Brazilian agriculture has been associated with deforestation since the 1970s. Between 1995 and 2005, average forest loss in Brazil averaged 20,000 square kilometers (or 2 million hectares) each year. As a result of deforestation, Brazil has lost 19% of the Amazon rainforest and 44% of the cerrado. Deforestation is a major source of greenhouse gas emissions and loss of biodiversity. Even with this loss, 65% of the Brazilian territory (about 550 million hectares) is still covered by native vegetation, which must be protected. The good news is that the deforestation rate has been significantly reduced since 2005 and now less than 5,000 square kilometers is cleared each year.

This trend suggests that Brazil may be near a turning point in effectively controlling deforestation and might eventually recover the lost ground in the future. Several factors help explain this reduction in deforestation rate and lend support to a more favorable outlook. The first one is less population pressure. As fertility rates drop and population migrates to urban areas, there is less pressure to clear land. Increased agricultural productivity also reduces deforestation as more food is grown on the same amount of agricultural land. For example, productivity gains in beef cattle production between 1990 and 2014 reduced land use demand by 250 million hectares.

Arguably, the major factors associated with the recent reductions in deforestation rates in the agricultural frontier are policy and better governance. Beginning in the mid-1990s, the Brazilian government created a system of national parks and wildlife and indigenous reserves covering 44% of the Legal Amazon where farming is banned. The federal environmental police (called Ibama) was beefed up to enforce the law. Law enforcement is now made easier with remote sensing technology showing in real time where environmental violations occur. Imazon, a Brazilian NGO, uses NASA satellite data to track deforestation and a platform developed by Google to process the data quickly and publish them on a monthly basis. The government published a list of municipalities with the worst environmental records in 2008 and withheld subsidized credit from them. The press embraced the environmental agenda, publishing deforestation data to society and putting pressure on politicians and industry participants to respond. Burning down the rainforest is not only outlawed but socially unacceptable.

Under pressure from NGOs and the press, industry reacted with the adoption of sustainability policies and better governance. We presented the sustainability efforts of UNICA in the sugarcane industry (Chapter 4) and Aprosoja in Mato Grosso (Chapter 5). Other noteworthy examples include moratoriums on purchases of soybeans and cattle produced on illegally cleared land. Multistakeholder initiatives such as Bonsucro (sugarcane), the Soybean Working Group (GTS), and the Sustainable Livestock Working Group (GTPS) bring producers, processors, NGOs, and government agencies together to find solutions to environmental problems and certify sustainable practices.

Another major policy change was the passage into law of the new Forest Code (Código Florestal) in 2012. The centerpieces of the Forest Code are the rural environmental registry (Cadastro Ambiental Rural — CAR) and the environmental compliance program (Programa de Regularização Ambiental — PRA). All 5.2 million farmers in Brazil have to file the CAR by May 2016, describing their environmental assets and liabilities. The CAR reduces uncertainty about land use because it shows a clear picture of who is using the land and how much forest it is supposed to have. In doing so, it will be easier to enforce environmental laws and control deforestation. Farmers without a CAR will have an illegal status. Banks will require loan applicants to show a CAR; traders, processors, and supermarkets will buy only from producers with a CAR.

But the CAR is just the first step. The CAR shows whether a producer falls short in complying with environmental laws. All producers with any environmental liability will not be able to clear more land and will have to apply for an environmental compliance plan (the PRA) to replant native trees or allow for the natural regeneration of forest in illegally cleared land. It is estimated that 20—25 million hectares of forests will be replanted in the existing farm establishments. It is still uncertain how much forest restoration will cost and who is going to foot the bill for the PRA, but progress in conservation and environmental compliance is clearly being made.

6.3.2 Social Issues

In addition to environmental problems, agricultural development in Brazil has also been associated with social concerns, including labor conditions and smallholder exclusion. Historically, labor conditions in agriculture have been neglected. Similarly to the environmental concerns discussed above, NGOs and the press started to call to the attention of the Brazilian society the labor issues in agriculture and agribusiness, including slave and child labor. Society, in turn, put pressure on politicians and industry participants to address these concerns.

In 2005 the Brazilian government introduced NR-31 to regulate worker safety and health in agriculture. NR-31 includes 259 rules to be followed by farmers and agribusinesses employing rural workers. These rules include the definition of proper work conditions and measures to reduce risks to worker safety and health. Enforcement of the law is carried out by labor inspection agents who particularly target large commercial farms and agribusinesses. Farm employers who do not comply with these rules or are deemed by the inspection agent to put workers in hazardous work conditions face stiff penalties and risk having their names included in the "dirty list" of the Ministry of Labor.

As a response to more stringent laws and societal pressures, industry associations have responded with labor compliance and corporate social responsibility (CSR) initiatives. For example, Chapter 4 discussed the CSR agenda introduced by UNICA in the sugarcane industry and the signing of the National Labor Commitment. Chapter 5 presented the *Soja Plus* project led by Aprosoja in Mato Grosso to educate producers about the NR-31 rules and assist them in complying with labor and environmental laws.

The other social challenge associated with agricultural development is smallholder exclusion. As Brazilian agriculture developed with increasing levels of efficiency, it went through a structural change process that included specialization, consolidation, globalization, and the emergence of tightly coordinated supply chains. Productivity gains at the farm level and structural changes at pre- and post-farm-gate stages of the agrifood system led to consolidation at the farm level. We discussed in Chapter 5 how the crises of 1995 and 2005 forced many farmers to sell their land to pay off debt. Entrepreneurship in the agricultural frontier – 2,000 km away from the ports, with poor infrastructure and several market failures – is a risky business. We celebrate the ones who were able to make it, but we cannot forget that many more did not succeed. Not surprisingly, the number of farm establishments with more than 100 ha decreased by 10% between the last two censuses (1995/1996 and 2006). Consolidation occurs as a result of competition as the market works as a selection mechanism. Unfortunately, the current policies affecting commercial farms under the Ministry of Agriculture, Livestock, and Food Supply (MAPA) do not provide an adequate safety net to reduce the stress and hardship associated with intense competition. If the policy status quo persists, I suspect we will continue to witness a "disappearing middle" in Brazilian agriculture and the resulting emergence of a bipolar distribution of a small number of very large commercial and corporate farms coexisting with a large number of smallholders.

Data from the 2006 Census of Agriculture show that 500,000 of farm establishments, or 11% of the total, produced 87% of the total agricultural production value. On the other end of the distribution, 2.9 million farm establishments (two-thirds of the total) produced 3% of the total production value. Based on these numbers alone, Brazil is no different from the United States and the European Union, where 87% of total production occurs on 10% and 14% of farms, respectively. The major difference is that smallholders in developed countries often find work off-farm and the largest share of their family income is from employment in other sectors of the economy. In Brazil, however, smallholders are poor and generate a meager gross income of less than two minimum salaries per month. About 60% of these poor smallholders are in the northeast region, distant from urban markets, with poor infrastructure, low levels of education, and imperfect access to credit and technical assistance. The data lend support to the main thesis of this book, that without entrepreneurship at the farm level and without organization of well-organized supply chains, agricultural development will not occur. Farmers without access to basic factors of production (technology, credit, modern farm inputs) — either because of low capacity to absorb knowledge and new technologies or due to market and policy failures — and access to markets, did not participate in the agricultural transformation process and were left behind in poverty.

Unfortunately, there is a widespread belief in Brazil that agricultural development — and agribusiness in particular — is not inclusive and con-centrates income in a few rich people. The concept of agribusiness, how-ever, does not exclude smallholders. Coined by John Davis and Ray Goldberg from the Harvard Business School in 1957, the concept of agri-business includes all stages of the agrifood supply chain and suggests that effective coordination between all the participants involved in a particular value chain is the key to efficiency gains, technological improvement, and equitable value share. As shown in Chapter 3, for example, smallholders in the southern region are not excluded from agricultural development and are able to earn a fairly decent income from farming, much higher than the per capita national income, precisely because they are connected to well-organized value chains, led by either cooperatives or contract farming arrangements. The challenge is how to include more small-holders to well-functioning agrifood chains in the future. The examples provided in this book suggest that this process is endogenous and highly dependent on the collective entrepreneurship of farmers.

Policy responses to the challenges faced by smallholders in Brazil include land reform and the family farm program (PRONAF), both over-seen by the Ministry of Agrarian Development (MDA). Between 1995 and

2010, the land reform program settled more than 1 million families across the country. PRONAF provides subsidized credit — with much lower interest rates than the official rural credit lines for commercial farmers — and extension services targeted to smallholders. Federal law 11.326/2006 dictates that only very small farms (up to four fiscal modules), which employ primarily family labor and generate income predominantly from on-farm activities, are eligible for PRONAF subsidies and assistance. Even with this narrow definition of family farm, it includes about 4.4 million farm establishments — 84% of the total number of farms and 24% of the total agricultural area — based on data from the 2006 Census of Agriculture. Extension services focus on technologies adapted to small-scale agriculture including agro-ecological production systems. Family farmers also have preferential access to public schools and other institutional markets, organized by the government under the Food Acquisition Program (Programa de Aquisição de Alimentos — PAA). Regular income transfers to poor families under the *Bolsa Família* program and the increased coverage of social security and retirement benefits also contributed to a reduction in rural poverty and decreased incentives for urban migration. Possibly as a result of these policies, the number of smallholder farms with less than 10 hectares increased between the two latest censuses (1995 and 2006). The number of rural workers employed in subsistence farming also increased since the 1990s. The economic impacts of these policies are not well known. The low education level of the rural poor limits the impact of policies that attempt to increase productivity, provide access to markets, and increase smallholder income.

To conclude, despite the historical bad record in both environmental and social accounts, progress is being made. I am an optimist that future agricultural development in Brazil will be more environmentally sound and inclusive. But I do not believe in a revolution. Sustainable agricultural development will likely happen in a "modest, slow, molecular, and definitive" way.

INDEX

Note: Page references followed by "*f*" and "*t*" refer to figures and tables, respectively.

A

Agrária, 62–64
Agrarian organization programs, 36
Agribusiness, 157
 vertically integrated, 73
 orange juice, 100–107
 sugarcane, 77–100
Agricultural cooperatives, 45
 diversity of agricultural organization, 47–50
 Paraná, agriculture in, 50–65
 Santa Catarina, agriculture in, 65–70
Agricultural policy, 28–39
 evolution of, 29*t*
 government support, 33–39
 public investments, 28–32
 subsidized rural credit, 32–33
Agrifirma Brasil Agropecuária S.A., 129, 131–133
Agrinvest, 129
Agroconsult, 126
Agro-ecological zoning policy, 77
Agro-pastoral rotation system, 125
Agrosilvopastoral (ASP) systems, 121–122
AMPA, 135, 149–150
Aprosoja, 135–137, 139, 149–150, 156
Asian soybean rust, 117–118

B

Beef cattle production, 114–115, 122, 154
Biological Nitrogen Fixation (BNF), 25–26
Bolsa Família program, 157–158
Bonsucro, 154–155
BrasilAgro, 129–131
Brazilian Agricultural Research Corporation. *See* EMBRAPA
Brazilian Cooperative Organization (OCB), 51–56
Brazilian Sugarcane Industry Association (UNICA), 84–91, 149–150, 154–156

communication efforts, 90
 governance structure, 85–86
 organizational structure, 85–86
 sustainability initiatives, 86–89
Brazilian Technical Assistance and Rural Extension Corporation (EMBRATER), 32

C

Cadastro Ambiental Rural (CAR), 155
Campinas Agronomic Institute (IAC), 22–23
Campos Gerais, 58–59
Canarana, 137–138
Castrolanda, 58–62, 153
Cattle ranching, in Mato Grosso, 113–115
Ceagro-Mitsubishi, 129
Central cooperatives, 65–67
Cerrado, 111
 development of, 27–28
 crop production in Brazil and Mato Grosso, 115–116, 115*t*
 entrepreneurship, 122–134
 commercial producers, 124–129
 corporate farms, 129–133
 economies of scale, 133–134
 evolution of farming in Mato Grosso, 112–116, 115*t*
 producer organizations, 134–141
 Aprosoja, 135–137
 new-generation cooperatives, 137–141
 productivity gains, 116–122
 EMBRAPA agrossilvipastoril, 121–122
 Mato Grosso Agricultural Research Foundation, 118–121
 soil, 20–21
Citrosuco, 101–103, 106–107, 148
Colonizadora Nova Eldorado, 125

159

Printed in the United States
By Bookmasters